Workbench Guide To Semiconductor Circuits and Projects

Workbench Guide To Semiconductor Circuits and Projects

Michael Gannon

PRENTICE-HALL, INC.
ENGLEWOOD CLIFFS, NEW JERSEY

Prentice-Hall International, Inc., *London*
Prentice-Hall of Australia, Pty. Ltd., *Sydney*
Prentice-Hall of Canada, Ltd., *Toronto*
Prentice-Hall of India Private Ltd., *New Delhi*
Prentice-Hall of Japan, Inc., *Tokyo*
Prentice-Hall of Southeast Asia Pte. Ltd., *Singapore*
Whitehall Books, Ltd., *Wellington, New Zealand*

© 1982 *by*

Prentice-Hall, Inc.
Englewood Cliffs, N.J.

All rights reserved. No part of this book may be reproduced in any form or by any means, without permission in writing from the publisher.

Library of Congress Cataloging in Publication Data

Gannon, Michael,
  Workbench guide to semiconductor circuits and projects.

  Includes index.
  1. Semiconductors—Amateurs' manuals. I. Title.
TK9965.G26   621.3815'2   81-18880
                              AACR2

ISBN 0-13-965277-9

Printed in the United States of America

This Book's Practical Value To You

In the chapters that follow, you'll find illustrated plans for a wide assortment of semiconductor circuits and projects that will pay for themselves many times over through useful service. If you're like most of my colleagues in the field, you'll discover just the kind of clear instructions and practical circuits you're after. Each circuit serves a worthwhile purpose around the home or workplace... can be constructed of easily available parts... and is relatively simple in design. As an added bonus, many can be completed in just a few hours of enjoyable, productive effort.

In order to provide the maximum amount of information that you need with each circuit diagram, power supplies are grouped together in the last chapter and referred to by number. For easy reference during assembly, semiconductor pin assignments have been included directly on each diagram and parts are detailed in a list which immediately follows. As a means of further clarifying technical points mentioned on the diagrams and in the text, schematic symbols and other related matters are explained in a special section.

You'll want to read Chapter 1 carefully before you proceed to the circuits and projects that follow. This chapter contains helpful suggestions, timesaving short cuts, and a quick review of construction techniques that will make your efforts more productive. It also provides information that will show you how to secure low-cost parts and supplies as well as inexpensive (if not free) data sheets, application notes, catalogs, and manuals. As a practical matter, you can test many of the circuits which you intend to use in less than 30 minutes with the easy breadboarding techniques explained in Chapter 1. Here are just a few examples of the valuable circuitry you will find in succeeding chapters ...

Wherever you live, you'd like to make your surroundings more energy-efficient without sacrificing the comfort and convenience we've come to regard as normal. You can slash energy consumption with the power-savers described in Chapter 2. You'll find such ideas as the Light Miser, which automatically regulates energy flow to an incandescent bulb in order to provide just the right amount of illumination for your purpose regardless of changes in ambient light conditions; or the Pushbutton Switch that "Remembers," which automatically shuts OFF an appliance after a pre-selected interval, thereby directly reducing not only your electric bill, but possibly your heating and air conditioning bills as well.

Chapter 3 contains successful formulas for putting inexpensive semiconductors to work for you around your home or apartment in creative, practical ways. If, like most technicians and experimenters, you want to know more about ways to use solar energy, don't miss Chapter 4. You'll find many circuits for harvesting nature's free and ultimate energy source. There's a solar-powered electric generator described, a sun-powered strobe light/beacon flasher, sun-tracking circuits, and much more.

Robots may have been a fantasy in the 50's, but they have become a reality in the 80's. They need not, of course, resemble the robots in movies ... but you can construct your own tireless fleet of helpers with the detailed instructions found in Chapter 5. They come in a variety of sizes and shapes, including one that will keep a liquid storage tank filled to within whatever limits you specify; or another that will close a relay for a timed interval at dusk (or dawn) to perform a task such as watering your lawn or turning ON and then OFF a sign at your place of work. Still another will sequentially operate a bank of relays to complete any job from developing film to mixing cement.

You will be able to explore inexpensively the very latest in integrated circuit sound chips with the Chapter 6 diagrams. This portion of the book is loaded with straightforward plans for producing pink noise generators, 2-way intercoms, and sound-boost units. It even contains plans for producing complete stereo systems, pre-amps, mixers, and a variety of other functional products.

In addition, there are projects that will help you add safety and convenience to your vehicle at a fraction of the dealer price. Chapter 7 instructions will show you how to complete such inexpensive electronic mechanisms as a backup sounder for alerting pedestrians in the immediate vicinity as you shift your vehicle into reverse. And to supplement the "idiot lights" on your dashboard, it's a simple matter to add a buzzer circuit which will sound in the event that any of these lights should come ON indicating trouble. It can be a costly matter if you fail to notice one of these lights on a bright day. This circuit could save you many times over the cost of this book.

We've also included a 12-volt to 120-volt inverter for operating home-type appliances in your vehicle, a 12-volt battery charger, a low brake-fluid monitor, a headlamp ON reminder— and many more useful projects that will improve your vehicle.

This Book's Practical Value to You

Other chapters are crammed with circuits and projects for your vehicle, home, and workshop. In summary, you'll find that this guide does indeed emphasize *practical* semiconductor circuits and projects. Each one is designed to save you time—and money—while increasing your level of enjoyment in the fascinating field of electronics.

Michael Gannon

CONTENTS

This Book's Practical Value to You 5

1. **How to Accomplish More with Semiconductors** 17
   How to Buy Semiconductors by the Generic System 20
   Access to Information on IC's 21
   Key Elements in an Efficient Workbench 22
   An Amazingly Easy Way to Breadboard 24
   Playing It Safe with High Voltage 27
   Less Heat Means Longer Life 28
   Top Performance at Minimal Cost 29
   How to Wire for Dependability 30
   How to Assemble a Circuit in 12 Easy Steps 31
   Helpful Schematic Symbols 41

2. **Semiconductor Circuits and Projects that Reduce Home Energy Consumption** 49
   High-Efficiency Dimmer/Motor Speed Control 52
   Light Miser 53
   Correct Power Factor for Amazing Energy Savings 54
   How to Stay Warm at Night and Still Reduce Your Fuel Bill 55
   Temperature-Sensing Fan Regulator 56
   Pushbutton Switch that "Remembers" 58
   Ambient Light Controller 60

3. **Using Semiconductors to Build Home Convenience Circuits and Projects** 63
   Battle Lantern 66
   Doorbell of Two Tones 67
   Rain Alarm 68
   T.V. Commercial Killer 70
   Carbon-Zinc Battery Refresher 71
   Automatic Pump Control Unit 72
   Electronic Scarecrow 74
   Electronic Flypaper 76

4. **Circuits for Harvesting the Free Energy of the Sun** .. **79**

   Solar Electric Generator 82
   Self-Powered Strobe Light/Beacon Flasher 83
   Pump Control for Solar Water Heater 84
   Solar Energy Trap 86
   Left-to-Right Tracker for Your Solar Array 89
   Return-to-Dawn Control for Your Solar Array 91
   Up/Down Control for Your Solar Array 93

5. **Ingenious Yet Simple Robot Circuits for Home, Shop, or Industry** **97**

   Precise-Temperature Water Heater 100
   Electric Switch with a Memory 101
   Tank-Filling Robot 103
   Relay Control that "Sees" 105
   Relay Control that "Hears" 107
   Tireless 1-2-1-2 Robot 109
   1-2-3 ... Robot 111

6. **How to Use Inexpensive Semiconductors for Top Quality Sound Reproduction** **115**

   Waterfall and Ocean Background Sounds on a Chip 118
   Handy 2-Way Intercom 119
   Extra Sound for Any Low-Powered Audio System 120
   PA/Megaphone 122
   Inexpensive High-Fidelity Stereo Phonograph/Tape Recorder 124
   Super Stereo Preamp 126
   Hi-Fi Magnetic Cartridge Preamp 128
   Provisional Stereo Mixer 130

7. **Practical Semiconductor Circuits for Your Car** **133**

   Back-up Sounder 136
   Handy 12-Volt to 120-Volt Inverter 137
   12-Volt Automotive Battery Charger 138
   Low Brake Fluid Monitor 140
   Idiot Light Audio Back-up 141
   Headlamp-ON Reminder 143
   Intrusion Alarm for Your Car, Truck or Van 145

8. **Semiconductor Projects for "Fun" and Profit** **147**

   CW Oscillator for Morse Code Practice 150
   Handy Magnetizer/Demagnetizer 151
   Beat Generator 152
   Electroplating Machine 153

Contents 13

    Lie Detector 155
    Sports Starting Light Sequencer 157
    Sound Microscope 158

9. **Dynamic Semiconductor Warning Mechanisms** 161
    Building Silent Sentries 163
    Chemicals Cabinet Alarm 164
    High-Water Alarm 166
    High-Temperature Alarm 167
    Low-Temperature Alarm 169
    Break-beam Alarm 172
    Complete Home Alarm System 173
    Touch Alarm (for use with Timer Circuit) 176
    Light-Detecting Alarm (for use with Timer Circuit) 178
    General Purpose Timer 179

10. **Using Semiconductor Circuits for Testing and Measurement Purposes at Your Workbench** 183
    Hi/Low Logic Tester 186
    Capacitance Box 187
    Inline Current Limiter 188
    Buzz Box 189
    Variable-Frequency Square-Wave Generator 190
    Audio-Signal Generator 192
    Capacitance Tester 193

11. **Semiconductor Power Supplies and Auxiliary Circuits 197**
    INTEGRATED CIRCUIT-REGULATED SUPPLIES
    5-Volt, Low-Current 200
    5-Volt, Medium Current 201
    5-Volt, High Current 202
    9-Volt to 12-Volt Adjustable, Low-Current 204
    9-Volt to 12-Volt Adjustable Medium-Current 205
    15-Volt, Low Current 207
    12-Volt to 15-Volt Adjustable, High-Current 208
    ±9 Volt to ±12 Volt, Dual Complementary 210
    15 Volts to 20 Volts, Dual Complementary 212

    ZENER DIODE-REGULATED SUPPLIES
    9-Volt, Low Current 215
    12-Volt, Low-Current 217
    15-Volt, Low-Current 218
    1.5-Volt to 9-Volt Adjustable, Low-Current 219

±9 Volt, Dual Complementary, Low-Current 221
15 Volt, Dual Complementary, Low Current 223
5.25-Volt, Battery-Powered Supply for TTL Circuitry 225
9-Volt, Low-Current, Battery-Powered 226
±9 Volt, Dual Complementary, Low-Current, Battery-Powered 227
12-Volt, High-Current, Battery-Powered 228

DECOUPLER/REGULATOR FOR USE WITH MOBILE PROJECTS

Decoupler/Regulator 229

Index **231**

Workbench Guide To Semiconductor Circuits and Projects

How to Accomplish More with Semiconductors

The use of amazingly functional yet low-cost semiconductor circuits is expected to proliferate in the home, workplace, school and in vehicles, making our lives not only easier, but more enjoyable as well. And whether you pursue this rapidly expanding field as a full-time profession or merely as a hobby, your satisfaction of accomplishment and material reward should be most heartening.

Looking backwards for a moment, most semiconductor devices that we use today were born and came of age in the 50's and 60's. The discrete transistors, diodes, and other solid-state devices developed during these decades were not only smaller, longer lasting, and more energy-efficient than the vacuum tube devices which they replaced, but considerably less expensive as well.

But it was during the 70's that the integrated circuit—which may have an impact on society comparable to that of the wheel—was perfected. Here, entire semiconductor circuits, which previously had to be laboriously assembled of discrete components, could now be reproduced in miniature form on a silicon surface using a photographic process. And with the addition of only a few parts to bias the IC (or because not all components can be produced with these photographic techniques), a complex unit of extraordinarily small size and low cost was the result.

Of course this continuing trend toward large-scale integration of components onto a single silicon chip necessitates a certain change in attitude on the part of the technician. Whereas we previously had to be concerned with the dozens of parts which made up, for example, an audio amplifier—this entire unit (minus a few biasing components) is now available in the form of an integrated circuit about the size of a cockroach. IC's which perform entire block functions now snap together as discrete components once did. And actually, except for the purpose of designing circuits or for academic considerations, there's little reason to even wonder what's inside of them.

But now it's time to get down to business. In order to help you construct the circuits and projects of succeeding chapters, what you'll want first is information on how to secure components and technical information with the least possible effort and expense. Then we'll follow up with a review of techniques and practices which will make your time more productive. And finally, a hands-on picture demonstration showing recommended ways to assemble a typical circuit found in this guide.

HOW TO BUY SEMICONDUCTORS BY THE GENERIC SYSTEM

Despite the alpha-numeric soup which manufacturer and dealer identification codes for semiconductors seem to make, there *is* an industry-wide coding system in widespread use. And to save you time and expense in securing the semiconductors that you will need for completion of the circuits and projects of following pages, virtually all of the called-for components adhere to this "generic" system. Here are some guidelines that will help you identify diodes, transistors, and IC's by the generic number which is usually stamped on each piece.

In this standardized system, diodes (including zener diodes) follow the alpha-numeric sequence 1NXXXX. For example, 1N4001 refers to a 1-amp, 50-PIV diode which is produced by a variety of manufacturer's under this same code. As for discrete transistors, most follow the alpha-numeric sequence 2NXXXX, although there are numerous exceptions-to-the-rule here. Most integrated circuits have a three-to-seven digit code which identifies the type of circuitry contained within. However, in the case of integrated circuits some further explanation is in order.

The generic code stamped on each IC is almost always preceded with and followed by "prefix" and "suffix" lettering codes specific to a particular manufacturer. For example, three functionally identical "555" timing chips might be labeled: NE555P, LM555, and LM555C. In the latter two instances, the prefix "LM" identifies the manufacturer as National Semiconductor Corporation ... and the LM555 has an operating temperature range of from minus 55°C to plus 125°C versus 0°C to plus 70°C for the LM555C. In some instances it would be necessary to select a particular grade of device to suit a specific application. But more on this later.

For obtaining the semiconductors and other components called for in the following pages, virtually all are off-the-shelf items at any electronics supply store. Since Radio Shack stores are the largest retail outlet, Radio Shack catalog numbers for the needed parts (or for ones which will substitute), where applicable, are also included directly on the parts lists. To cut costs even further, consider ordering components through listings which appear in the back pages of many amateur electronic magazines such as *Popular Electronics*. And to secure components for next-

How to Accomplish More with Semiconductors 21

to-nothing, a scrap chassis which you may already have or can buy very cheaply from an electronic repair shop can be a veritable gold mine.

Of course these recommendations and all other recommendations found in this guide are intended solely for the benefit of you, the reader. There are absolutely no financial or other interests involved.

ACCESS TO INFORMATION ON IC's

There's always a need for information on integrated circuits. As previously mentioned, sometimes it is necessary to select a particular grade of device to suit a specific application, or data may be required for any number of other reasons.

For any integrated circuit information search, the IC MASTER is often the best place to start. It catalogs virtually every IC made by generic code ... cross-references prefix and suffix codes with the appropriate manufacturer ... and even includes key electric parameters for many devices which may suffice to answer your questions. It also lists the addresses of manufacturers where you can write for more information. However, with over 45,000 IC devices being manufactured at the time of this writing, the IC MASTER has become a multi-volume text more suitable for a technical library than a technician's workbench. You can obtain ordering information at the following address:

    IC MASTER
    UPT/Hearst Business Communications, Inc.
    645 Stewart Avenue
    Garden City, N.Y. 11530
    Attn: Circulation
    Telephone (516) 222-2500

And don't overlook the manufacturers themselves as a source of data on IC's. For information on a particular device type, they will often send you a free data sheet. Or for information of a more general nature, ask them for their free application notes (which contain suggestions for using their new products) data books (which detail the electric parameters of a particular class of device) and catalogs. Of course, best service is usually forthcoming if you can use stationery which has a company heading.

For technicians engaged in electronics as a profession or interested in designing their own circuits, some of the larger companies publish excellent data libraries. Available texts include data books, circuit design handbooks, and various reference manuals. One good source of this material is National Semiconductor Corporation. You can obtain an ordering envelope complete with a description of texts and their current prices at the following address:

> National Semiconductor
> P.O. Box 60876
> Sunnyvale, CA 94088
> Telephone (408) 737-5000

Another good source is:

> Texas Instruments Incorporated
> P.O. Box 225012, Mail Station 54
> Dallas, Texas 75265
> Telephone (214) 238-4844

KEY ELEMENTS IN AN EFFICIENT WORKBENCH

The old adage that a craftsman is only as good as his tools applies as much to the field of electronics as to any other. Of course, tools and equipment are no substitute for skills and knowledge. Still, an efficient workbench setup is almost mandatory for best results. Some key points to keep in mind here are comfort, good lighting, and adequate tools and test instruments.

Comfort is of first-line importance. As shown in Illustration 1-1, a very serviceable arrangement can be quite inexpensive. A chair which gives proper body support allows you to work long, involved hours without ending the day with an aching back or kinked neck. An old office chair can be wonderful for this purpose. On this same line, you'll need a solid work surface which is at just the right height. Here, a surplus office desk fills all the requirement and provides plenty of drawers for handy storage of parts and tools. Also important is a quiet location such as the corner of a basement, and a stereo music system.

And don't overlook good lighting. For example, the fluorescent-type light shown gives bright, low-shadow illumi-

How to Accomplish More with Semiconductors 23

Illustration 1-1. Three key points in setting up your workbench are comfort, good lighting, and adequate tools.

nation with a minimum of unwanted heat. A swivel mount allows the fixture to be positioned close-in for delicate operations ... or further away for normal use.

As a general rule, adequate tools and test instruments can cost as much or as little as you have to spend for them. On the lower end of this scale, a somewhat obsolete yet completely functionally oscilloscope (Illustration 1-1) can usually be bought at a surprisingly low price from a used electronic equipment supply store (check the Yellow Pages of your phone directory). And you can construct a wide variety of test instruments plus bench-type regulated power supplies using the schematics found in Chapters 10 and 11. Illustration 1-2 shows power supply 11-8 and 11-14 (see Chapter 11) constructed for use as a bench unit.

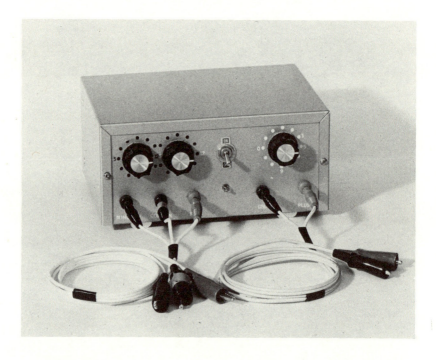

Illustration 1-2. Power Supply 11-8 and 11-14 as used for a bench unit

AN AMAZINGLY EASY WAY TO BREADBOARD

Sooner or later most electronic technicians come to the same conclusion: It's best first to breadboard a new circuit before attempting to construct it in permanent form. One reason for this is that it allows you to pre-test the circuit quickly in order to become familiar with its operation. Another is that it lets you test the parts which you intend to use, discarding those which are defective or weak. Of course, experimenters often breadboard circuits as an end in itself.

Illustration 1-3 shows an extraordinarily quick and easy way to pre-test virtually any circuit found in this guide in less than 30 minutes. Here, strips of Velcro™ are glued or tacked in rows across a spacious assembly board (Velcro is a stick-together fabric available at most sewing supply stores). Next, components are attached to Velcro-backed wooden slats ... positioned on the assembly board in the same order as seen on a particular schematic diagram ... and then interconnected with jumper leads.

Illustration 1-3. An amazingly easy way to breadboard

For a few improvements to this design, heavy copper wire attached to the upper and lower extremes of the assembly board serves as positive and negative buses for quick attachment of

26   *How to Accomplish More with Semiconductors*

jumper leads. And painting the face of the board white improves "seeability" by providing a visual background to throw the components into relief.

For mounting the components on the wooden slats, attachment may be either temporary or permanent. For temporary attachment (Illustration 1-4), alligator clips snap to component leads, and small brass nails provide attachment points for the jumper leads. For permanent attachment, such as for making a breadboarding kit (Illustration 1-5), component leads are simply soldered to the brass nails. This picture also shows how to prepare a socket for the purpose of breadboarding an IC.

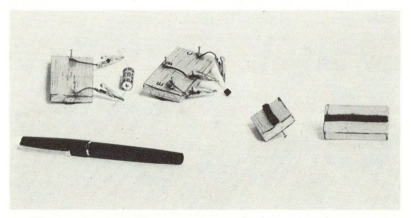

Illustration 1-4. How to temporarily mount components on wooden slats

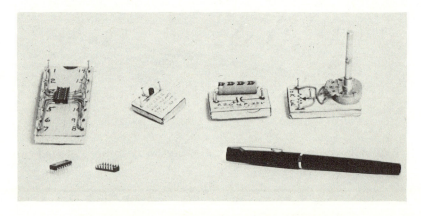

Illustration 1-5. Breadboarding an IC, and permanently-mounted components

How to Accomplish More with Semiconductors 27

PLAYING IT SAFE WITH HIGH VOLTAGE

When you are assembling a line-powered circuit, always remember that potentially hazardous high voltage will be present and take the necessary precautions. Figure 1-1 illustrates some of the safety precautions that you *must* take. It also shows some of the finer points of circuit construction that we shall be discussing presently.

As you'll notice, the transformer is electrically isolated from any conductive object that you might touch, and all line connections to the ON/OFF switch and to the transformer are securely fastened and well-insulated. One good way to fasten/insulate joints is with plastic-insulated connectors of the screw-on or snap-on variety. Another way is by soldering them directly to the circuit board in such a fashion that they can neither touch any other component nor be accidently pulled out.

Also, notice that all line connections are made inside the enclosure box. This serves as double protection for, in the event of a high voltage short circuit, the resultant sparks and flames will likely be contained within the box.

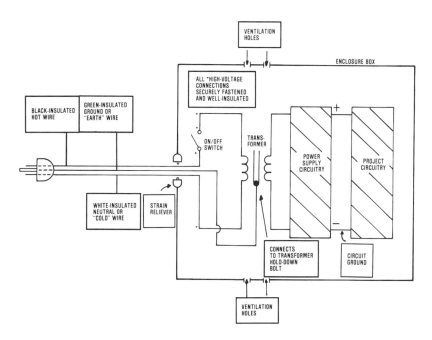

Figure 1-1. Safety precautions you *must* take, plus assorted tips

On this same topic, a chafe guard is mandatory at the vulnerable point where a line cord enters a metallic enclosure box. A strain reliever of the type used on many commercial appliances is probably the most effective piece of hardware for this purpose as it completely immobilizes the cord at this point. As an added advantage, it also prevents the line cord from being pulled out of the chassis if it is accidentally given a hard yank.

Although the use of a line cord with a third-wire ground is optional, it's cheap insurance against mishap in a number of ways. As a quick recap, the non-current carrying ground or "earth" wire of your house wiring system (green-insulated or else bare) as well as the current-carrying neutral or "cold" wire (white-insulated) are both electrically connected to either a metal rod driven into the ground or else to a water pipe. Therefore, both of these wires are maintained at zero volts potential and are unlikely to give you a shock when touched (as will the black-insulated hot wire which carries both current and voltage). But since the negative wire from a dc power supply is also referred to as "ground," when in reality there is not necessarily any electrical connection between the two, henceforth, let's call the third-wire of your house system the "earth" wire to maintain this important distinction.

Now, one of the safety features of using a three-wire line cord is that the three-pronged plug can only be inserted into a wall receptacle in one way—not two ways as with most two-pronged plugs (still referring to Figure 1-1). This ensures that the black-insulated wire of your line cord is connected to the hot side of your house supply; the white to neutral; and the green to earth. This means that when you open the ON/OFF switch to shut down the circuitry, the transformer truly goes to zero volts potential. This eliminates the possible source of leakage or shock.

As another safety feature, the earth wire can be used as a "fail safe" to guard against hazardous shocks. Attached to the transformer core, it prevents the case of this component from becoming a shock hazard in the event of insulation failure within. As double protection, the earth wire can also be connected to a metallic enclosure box thereby maintaining it at a safe zero volts no matter what happens within.

LESS HEAT MEANS LONGER LIFE

Taking steps to ensure adequate cooling of circuit components will pay you dividends by prolonging their service life. As illustrated (Figure 1-1, again), ventilation holes in the top and bottom of the circuit enclosure box allow convection currents to carry away excess heat from parts. Especially effective are holes located directly over and under such heat-generating units as power transformers and heat sinks.

If top-and-bottom ventilation holes are unpractical for some reason, side ventilation holes can be almost as effective. If you use side holes, however, be sure to locate them along the upper and lower portions of the box in order to promote the best circulation of cooling air.

Another way to promote longer parts life through heat reduction is by thoughtful placement and mounting of components on the perfboard. Components which are vulnerable to heat deterioration, notably electrolytic capacitors, are best located at least one inch away from such heat generators as power transformers and heat sinks. And for power resistors, zener diodes, and modular bridge rectifiers, mounting the component slightly above the perfboard (about ¼-inch) will permit cooling air to circulate underneath.

TOP PERFORMANCE AT MINIMAL COST

As a few additional time and money-saving tips, although the components called for in the circuit diagrams of succeeding chapters are listed by their correct values, sources of supply and availability of parts may differ widely. As a general rule, if your dealer doesn't carry a part of the specified voltage or power dissipation rating, a unit with the next higher rating will suffice. For example, if a diagram calls for a 2-amp, 100-PIV modular bridge rectifier and your choice is between a 1.4-amp 100-PIV unit and a 4-amp 100-PIV unit, the heavier rectifier is the better selection. The difference in price is usually minimal, and installing an undersized component can be very poor economy.

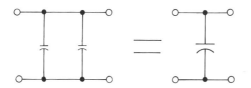

Figure 1-2. Smaller capacitors of the same type may be shunted together to yield a larger value.

For substituting capacitors, a unit made of a different material than specified or of a significantly higher voltage rating than what is called for (more than four times) cannot usually be interchanged with good results. However, if a capacitor of the specified number of microfarads is not available, smaller capacitors of the same type may generally be shunted together to yield the larger value. This is illustrated in Figure 1-2.

HOW TO WIRE FOR DEPENDABILITY

Although fuses are a bother to install and sometimes seem to blow for no reason at all, they can be a necessary inconvenience. Even though the IC voltage regulators used in many power supplies incorporate internal current limiting, their shut-down point may be too high to protect weaker power supply components from damage in the event of a short circuit (such as the smaller varieties of power transformer and bridge rectifier). Compared to many other components, a fuse is an easy and inexpensive part to replace.

As a rule-of-thumb for selecting the correct size fuse, first choose one equal to or slightly below the current rating of the weakest component in the power supply. Then, if this fuse tends to open during times of high transient current demand (such as when a relay is first energized), try a slow-blow fuse of this same rating before going on to a fuse of a higher rating.

On another equally important topic, one of the main causes of false triggering in digital circuits and in alarm circuits is pickup of stray voltages in interconnecting wires which is generated via fields. When you must route wires over some distance (such as for remotely locating a sensing element), first of all try to stay clear of power cables. The inductive field surrounding such cables—especially when a sudden surge of current flows through them—will induce a voltage in nearby wiring even though all insulation is intact. Here, the amount of inductive coupling between the conductors occurs in direct proportion not only to their proximity, but to the distance in which they run in common as well.

As another means of thwarting unwanted pickup from various electric, magnetic, and electro-magnetic fields when interconnecting circuit components over some distance, use 2-conductor speaker wire and twist it along its entire length at the rate of about four turns per foot. Twisting is especially effective in reducing pickup from such relatively distant sources as transmitting antennas and lightning storms. And for maximum protection in particularly adverse conditions, use shielded microphone-type cable with the protective outer braid connected to the earth wire of your home wiring system (and to circuit ground if applicable). The earth tends to absorb troublesome field energy.

But the most common cause of false triggering in digital and alarm circuitry is interference which is fed into the circuit via the line cord. Here, transient-producing gear such as a fluorescent light or a motor on the same fuse or circuit breaker generates interference which can cause circuit components to activate at the wrong time. When this problem occurs, the easiest remedy is often to plug the affected circuit into an electrical outlet which is fed by a different fuse or circuit breaker. Barring this, you might try installing an ac line filter on the transient-producing gear.

HOW TO ASSEMBLE A CIRCUIT IN 12 EASY STEPS

You can often put together an amazingly functional yet low-cost semiconductor circuit in only a few hours of pleasant, productive effort. As a typical example, this IC-based stereo amplifier delivers high-fidelity performance yet snaps together with less than 30 off-the-shelf components. The key ingredient, of course, is the IC which houses almost all of the amplification circuitry in miniature form. (Illustration 1-6)

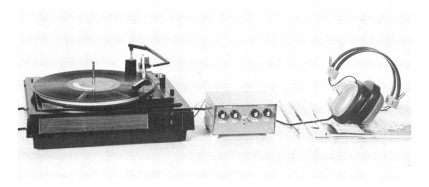

Illustration 1-6

Step 1. The first task in any semiconductor project is to gather all of the necessary components. For this purpose, a parts list is included with each circuit diagram. Here's a good tip: Be sure to select a perfboard—which serves as a non-conductive mounting board for the parts—of ample size in order to avoid crowding. Perfboard is available with various grid hole spacings and sizings, including a type suitable for direct insertion of DIP-packaged IC pins or IC socket pins. IC sockets (as shown) are recommended in order to avoid damaging the IC during the soldering procedure, and to make a possible IC replacement in the future a simple matter. Here's another good tip: Select single-strand hook-up wire for making quick connections between parts where no flexing or vibration will be present... multi-strand hook-up wire where flexing or vibration, however slight, will be present. Of course, use wire of sufficient current and insulation rating. For routing a weak signal between circuit components, use shielded cable with the protective outer braid connected to circuit ground in order to avoid unwanted pickup via fields. Also recommended is a shallow tray for holding small parts during circuit assembly to avoid their loss. (Illustration 1-7)

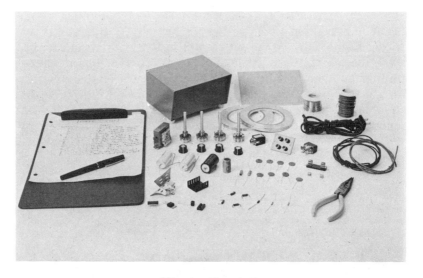

Illustration 1-7

Step 2. Before attaching the case of a semiconductor device to a heat sink (or chassis), be sure to apply a drop of silicon grease to their mating surfaces in order to promote heat flow. With the

TO-220 packaged semiconductor device as shown, silicon grease reduces case-to-sink heat resistance from about 1.2 for direct contact, to about 0.7 for a greased contact. However, if electrical isolation is required between case and sink, you must also include mica washers for insulation. Mica washers increase case-to-sink heat resistance to about 0.9 in this particular instance. (1-8)

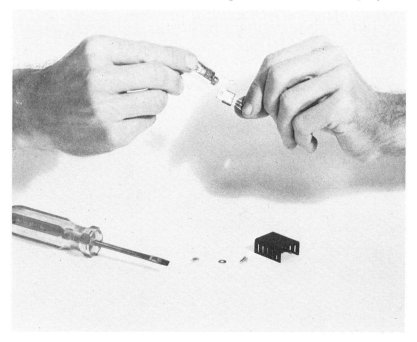

Illustration 1-8

Step 3. Mount components on the perfboard. Although there is no set method, some guidelines you'll want to keep in mind are as follows (Illustration 1-9):

- To ensure that all components will fit on the perfboard, trial-mount all components before soldering any into place.
- If possible, position all components in the same spacial relationship as they appear on the circuit diagram. This will greatly reduce the possibility of mistakes when making interconnections, and greatly speed up future troubleshooting tasks.
- Mount all components so that their identifying marks are clearly visible. This will make their later identification an easy matter.

- Install components which must dissipate considerable heat about ¼-inch above the perfboard to facilitate cooling. This guideline applies to power resistors, zener diodes, modular bridge rectifiers, and various other parts.
- If possible, mount electrolytic capacitors at least one inch away from heat sources such as power transformers and heat sinks. This will promote long-term reliability for these heat-sensitive capacitors.
- Keep component leads short, and position components as far away as possible from the power transformer. Short leads and ample spacing from the inductive field produced by the transformer reduce the possibility of hum pickup in audio circuits, and of unwanted triggering in digital circuits.

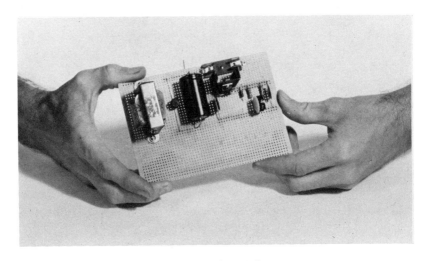

Illustration 1-9

Step 4. On the underside of the board, bend leads over in order to hold components in place for easy soldering. (1-10)

When a number of leads must be electrically interconnected, you can save time by busing them together. As shown, adhesive-backed copper foil, which is available at any stained-glass supply store, is affixed to the underside of the board and then punctured with a needle at each pre-punched hole prior to insertion of components. Leads may then be quickly row-soldered. Another good way to bus is with pre-tinned bus-type wire (such as Radio Shack 278-1341).

Some technicians prefer to make connections with wire-wrap techniques. Others favor printed circuit techniques.

How to Accomplish More with Semiconductors 35

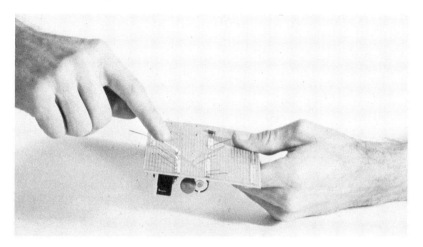

Illustration 1-10

Step 5. Proper soldering practices are of prime importance for first-time-through circuit success. Before applying solder, be sure that the joint is physically clean and mechanically tight. Soldering paste, of course, is not to be used on a perfboard as it leaves behind a conductive residue which can cause problems later. As a precaution against damaging semiconductors during soldering, attach an alligator clip to each lead up close to the component in order to absorb excess heat. Use a rosin-core 60/40 solder, and choose a thin diameter to facilitate placement of material and metering of quantity. Safety glasses are recommended during the soldering procedure to protect eyes from inadvertent splatters. (Illustration 1-11)

To apply solder, the idea is to heat *both* parts which are to be joined to the correct temperature, and then apply just enough material to fill the joint by capillary attraction. With too little heating, the molten solder has a grainy appearance and flows poorly—if at all—into the joint. With too much heating, plastic insulation in proximity to the joint begins to smoke and curl, and components may actually be damaged. With just the right amount of heating, the molten solder has a shiny appearance and soaks rapidly into the joint. But if too much solder is applied, the excess will run out, possibly bridging electrical pathways and causing a short. As a final tip, always allow the solder ample time to solidify before subjecting the connection to physical stress, or else a weak and grainy crystallization of the material will invariably result.

Step 6. Nip off excess lead lengths, and inspect each soldered connection. After donning a set of safety glasses, nip off each lead as close as possible to the joint. An old set of nail clippers

36 How to Accomplish More with Semiconductors

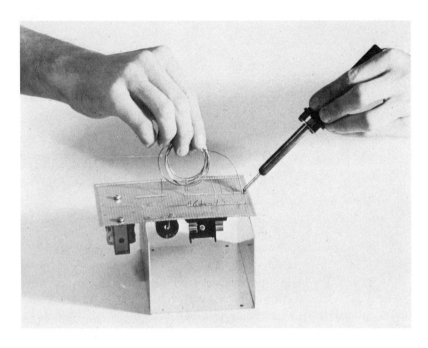

Illustration 1-11

is a handy tool for this task. Next, use a strong magnifying glass and bright illumination to inspect each solder connection for continuity, and for the unwanted presence of stray wire strands or solder drips which could cause a short circuit. Lastly, double-check to see that parts are interconnected in their proper relationships. (Illustration 1-12)

Step 7. Lay-out the face of the circuit enclosure box for installation of the potentiometers, input/output devices, etc., and then bore the necessary holes. A tall circuit box is to advantage here since it will allow these components to be fit above the area which the perfboard-mounted parts will occupy. This also makes later interconnections a simple matter. Some extra time spent in careful layout will usually save effort in the long run. Before drilling, use a centering punch to dent the spot where each boring must start to prevent the drill bit from wandering. (1-13)

Step 8. Secure the perfboard to the enclosure box. To do this, nut-and-bolt fastenings are usually the most practical means. First, hold the perfboard in position inside the enclosure box. Second, bore a hole clear through the bottom of the box and the perfboard, making sure to intersect the board at a spot which is

How to Accomplish More with Semiconductors 37

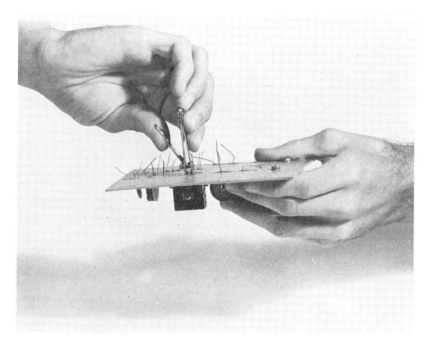

Illustration 1-12

Illustration 1-13

free of components. Next, use one nut to clamp the bolt to the box (as shown), and another two nuts in combination to clamp the bolt to the board. Repeat this procedure at four or five widely spaced locations around the board to form a rigid attachment.

A drop of white glue on nuts will ensure their tightness, yet allow fairly easy removal at some future time should this be necessary. (Illustration 1-14)

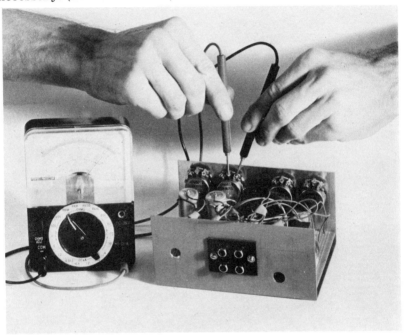

Illustration 1-14

Step 9. Make all of the necessary interconnections between the potentiometers, input/output devices, etc. and the circuit board components. For the potentiometers, an ohmmeter (shown) is a useful tool for determining their sense. The problem here is that the end terminals of potentiometers are often wired the opposite of what is expected in relation to the center tap. (1-15)

Step 10. Label all controls for easy identification. The most practical way to do this is with a letter-punching tool which crimps raised letters on a self-adhesive plastic tape. Another good way is with dry transfer decals (shown). To transfer a word, press the plastic sheet directly onto the circuit box where you want the word to appear. Then rub over the word with progressively firmer strokes of a ballpoint pen until the transfer is complete. (1-16)

Step 11. Holes drilled into the circuit enclosure box will contribute to the long-term reliability of the components inside by

How to Accomplish More with Semiconductors 39

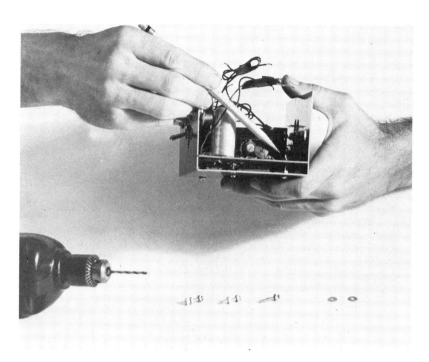

Illustration 1-15

Illustration 1-16

allowing excess heat to escape. Holes drilled into the top and bottom of the box—especially over and under such heat-

40 How to Accomplish More with Semiconductors

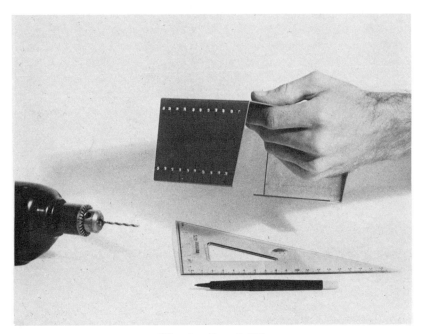

Illustration 1-17

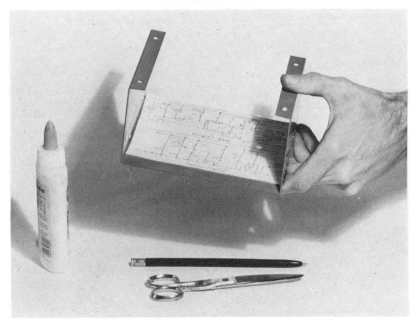

Illustration 1-18

How to Accomplish More with Semiconductors 41

generating components as power transformers and heat sinks—are most effective. However, if top-and-bottom holes are impractical for some reason, holes drilled along the sides toward the upper and lower extremes can be almost as effective. (1-17)

Step 12. As a final step, make a sketch of the circuit and paste it inside the circuit box cover. Be sure to show all component values, fuse sizes, and adjustment specifications. The technician who may have to service the set at some future date will appreciate this courtesy (and this technician will probably be you). (1-18)

HELPFUL SCHEMATIC SYMBOLS

	Battery, dry cell such as a standard carbon-zinc type
	Battery, multi-cell such as a lantern-type or an automotive-type
	Buzzer or bell. Note that they often have a polarity requirement, so check carefully.

42 How to Accomplish More with Semiconductors

	Capacitor, polarized such as most electrolytic types. Be sure to connect this capacitor into the circuit as shown on the schematic diagram or else it will have a short life.
	Capacitor, non-polarized such as a ceramic disk type. This capacitor can be connected into the circuit in either orientation.
	Ceramic phonograph cartridge (single channel).
	Diode.
	Diode, zener. Always double-check that you have this component connected into the circuit in the correct orientation to avoid damage when the circuit is energized.

How to Accomplish More with Semiconductors

	Diode, light-emitting. You will have to bias this component with a current-limiting resistor. For example, if the device is rated for 12 mA and circuit voltage is 12 V, you'll need a 1,000 ohm, ¼-watt resistor.
	Fuse holder and fuse. For an application where the fuse may blow frequently (such as a bench-type power supply), a panel-mount fuse holder is recommended. Otherwise an in-line or perfboard fuse holder should suffice.
	Integrated circuit. Note that pins are located so as to make each circuit diagram easy to read. The actual pin layout is detailed on a sketch accompanying each circuit diagram.
	Light bulb, incandescent.
	Motor. Remember that most types of dc motor are reversible. If you want it to turn the other way, simply reverse the connections.

44 How to Accomplish More with Semiconductors

(symbol)	Microphone
AC / Rect / AC (+) AC / diamond of diodes / AC (+)	Modular bridge rectifier. If you want to use discrete diodes for this purpose, connect them as shown:
(photoresistor symbol)	Photoresistor (resistance varies with light). There's almost never a polarity requirement, so you can connect this component into the circuit in either orientation.
WINDING / OFF / ON / CENTER TAP	Relay. When the winding is connected to power, the resultant magnetic field disconnects the center tap of the contact set from the OFF terminal, and connects it to the ON. Note that a relay may have from one to four contact sets. Be extra cautious when using relay contacts to switch line power. Always ensure that relay contact ratings are adequate for any appliance to which they may be connected.
—⟋⟍⟋⟍—	Resistor.

How to Accomplish More with Semiconductors 45

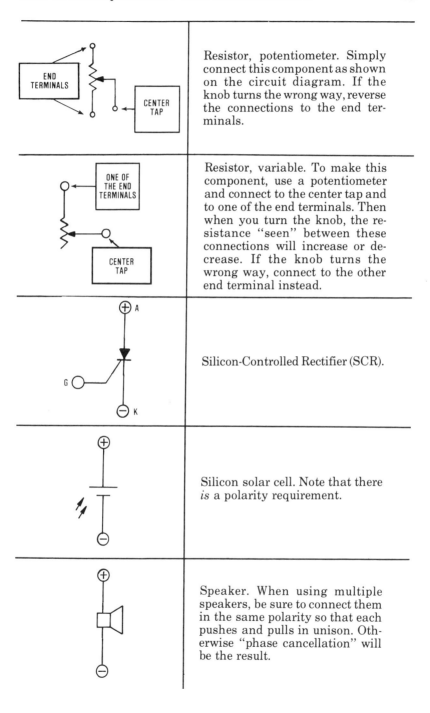

Resistor, potentiometer. Simply connect this component as shown on the circuit diagram. If the knob turns the wrong way, reverse the connections to the end terminals.

Resistor, variable. To make this component, use a potentiometer and connect to the center tap and to one of the end terminals. Then when you turn the knob, the resistance "seen" between these connections will increase or decrease. If the knob turns the wrong way, connect to the other end terminal instead.

Silicon-Controlled Rectifier (SCR).

Silicon solar cell. Note that there *is* a polarity requirement.

Speaker. When using multiple speakers, be sure to connect them in the same polarity so that each pushes and pulls in unison. Otherwise "phase cancellation" will be the result.

(symbol)	Switch, normally open (press to close).
(symbol)	Switch, normally closed (press to open).
(symbol)	Switch, ON/OFF (same as a single-pole, single-throw [SPST] switch)
(symbol with CENTER TAP)	Switch, single-pole dual-throw (SPDT). Clicking the switch connects the center tap to either one end terminal or the other.
(symbol with CENTER TAP)	Switch, rotary. Turning the knob connects the center tap to each of the multiple terminals in sequence.
(symbol with T)	Thermistor (resistance varies with temperature). Since there's almost never a polarity requirement, you can connect this component into the circuit in either orientation.

How to Accomplish More with Semiconductors 47

PRIMARY (120 VAC 60-Hz) / SECONDARY	Transformer. Be careful not to confuse the primary winding with the secondary winding or damage to the component will result.
FET symbol with G, D, S	Transistor. field-effect (FET), N-Channel.
PNP symbol with B, C, E	Transistor, PNP
NPN symbol with B, C, E	Transistor, NPN.
Triac symbol with MAIN TERMINAL 2, GATE, MAIN TERMINAL 1	Triac. Be cautious with this one as it carries high voltage.

2

Semiconductor Circuits and Projects that Reduce Home Energy Consumption

Although the cost of energy is likely to be on the upswing for the remainder of this century, higher prices do not have to mean a reduction in our standard of living. And semiconductor technology can play a significant role in making our surroundings more energy-efficient *without* sacrificing the comfort and convenience that we've come to regard as normal.

One semiconductor device that can be used to good advantage here is the Triac. It permits power to be throttled to an appliance with corresponding energy savings (and in many cases, appliances actually work *better* at a lower setting). Circuits 2-1 and 2-2 will show you some innovative ways to incorporate this device into practical, working circuits.

And don't overlook power factor correction to reduce your monthly power bill. Even that small amount of energy which "leaks" through the transformer primary of many circuits and appliances can add up into dollars over a month's time. Every little bit helps, and Circuit 2-3 will show you how to put an end to such unnecessary wastage.

But probably the most effective overall strategy to reduce home energy consumption is to turn OFF appliances when they are unneeded. But why run back and forth between switchboxes when semiconductor circuits can handle this task—automatically.

On this theme, you'll find such ideas as the Temperature-Sensing Fan Regulator (Circuit 2-5). When coupled to a summer ventilation fan, it will turn this fan OFF as the night air cools, thereby saving electrical energy. What's more, it can save you the trouble of getting up at night to accomplish this task.

Or what could be more convenient (or energy-efficient) than a circuit to automatically switch OFF an appliance such as a kitchen or bathroom exhaust fan after a pre-selected interval. The plans shown in Figure 2-6 will show you how to build such a unit.

As another energy-efficient idea, the 15%-20% of your electric bill that goes into lighting your home can probably be reduced by a noticeable amount with the installation of Ambient Light Controller Circuit 2-7. It will sequentially switch lights ON or OFF in a room to maintain that "just right" level of illumination. Convenience ... and energy savings.

Circuit 2-1. High-Efficiency Dimmer/Motor Speed Control

Energy savings plus improved service! This simplistic yet completely functional circuit allows you to operate appliances at reduced output with corresponding energy savings. As used with an incandescent lamp, it lets you choose the correct illumination for the occasion—from glimmer to full ON. It can even eliminate the need for a costly 3-way bulb. For a few more application ideas, you can use it to precisely adjust the output of a space heater, a hotplate, or any universal-type motor.

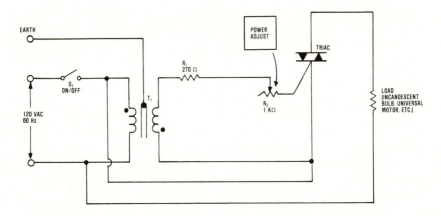

Circuit 2-1. High-efficiency dimmer/motor speed control

For construction, remember that high voltage will be present throughout most of this circuit, so take the necessary precautions. For increased utility, you can build this circuit into the base of a lamp or in an electrical wall box. Position "Power Adjust" variable resistor R_2 and ON/OFF switch S_1 for easy access from the front of the circuit enclosure. Also, notice that the ac wiring to the Triac is connected in reverse order to compensate for the phase inversion from transformer primary to secondary. Although almost all transformers are wound for a phase inversion, you can check yours with a dual-trace oscilloscope if you have any doubt. Be sure to select a Triac with the appropriate current rating for your intended application.

Semiconductor Circuits to Reduce Home Energy Use

PARTS LIST

R_1—270 ohm, ¼-watt (Radio Shack 271-1314)
R_2—1,000 ohm, ¼-watt (or more), linear taper potentiometer
S_1—On/OFF switch (such as Radio Shack 275-651)
T_1—6.3-volt secondary, miniature transformer (Radio Shack 273-1384)
Triac—Any Triac with sufficient current capability to drive load (such as Radio Shack 276-1001)

Circuit 2-2. Light Miser

A circuit which not only pinches your energy pennies, but automatically provides just the right amount of illumination for numerous applications around your home or shop. Simply energize this circuit, and set "Ambient Light Adjust" variable resistor R_2 for the desired illumination level. Henceforth, this unit will regulate energy flow to the incandescent bulb in order to maintain this level—from full ON to full OFF. Ideal for use in lighting staircases, hallways, and aquariums. You can even use it to control the main lighting source in a windowed room to provide energy savings during the brighter portions of the day. A truly remarkable circuit.

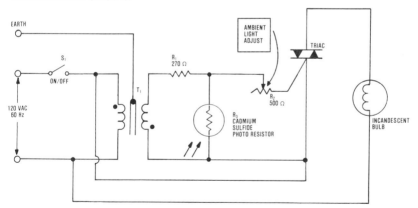

Circuit 2-2. Light miser

PARTS LIST

R_1—270 ohm, ½-watt (Radio Shack 271-016)
R_2—500 ohm, ¼-watt (or more) linear taper potentiometer
R_3—Cadmium sulfide photoresistor. About 500,000 ohms dark, 100 ohms light (Radio Shack 276-116)
S_1—ON/OFF switch (such as Radio Shack 275-651)
T_1—6.3-volt secondary, miniature transformer (Radio Shack 273-1384)
Triac—Any Triac with sufficient current capability to drive load (such as Radio Shack 276-1001)

54 Semiconductor Circuits to Reduce Home Energy Use

For construction, remember that high-voltage will be present throughout most of this circuit and to act accordingly. Be sure to position the R_3 photoresistor where it will receive a representative sampling of room ambient light. Position Ambient Light Adjust R_2 and ON/OFF switch S_1 for access from the front of the circuit enclosure. Note that the ac wiring to the Triac is connected in reverse order to compensate for the phase inversion from transformer primary to secondary. Although almost all transformers are wound for a phase inversion, if you have any doubt you can check yours with a dual-trace oscilloscope. Be sure to select a triac with the appropriate current rating for your intended application.

Circuit 2-3. Correct Power Factor for Amazing Energy Savings

Hard to believe, but you can reduce the power consumption of most line-powered circuits in this book by about 40% with the addition of a few inexpensive capacitors. Just connect your bench-type ac mA meter inline with the power cord—shunt the leads of this capacitance box across the transformer primary—and watch the current draw fall to a minimum as you switch in the right amount of capacitance through a trial-and-error procedure (then

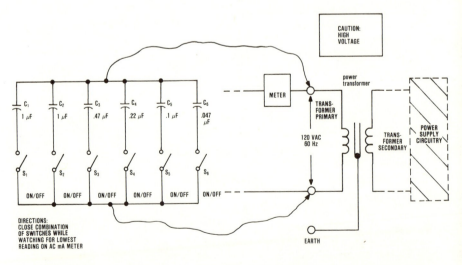

Circuit 2-3. Correct power factor for amazing power savings

Semiconductor Circuits to Reduce Home Energy Use 55

permanently solder this value of capacitance across the transformer primary). This has no effect on circuit performance except to greatly improve efficiency.

As a simplified explanation of power factor correction, a certain amount of power normally "leaks" through the primary winding of the transformer when it is connected to power. Adding the correct amount of capacitance in shunt with this winding "neutralizes" this leakage, resulting in improved efficiency.

For construction, remember that high voltage will be present throughout most of this circuit when in use, and act accordingly. Mount capacitors on a perfboard, with front-panel access for ON/OFF switches S_1 through S_6. Use insulated clips for connecting this circuit across the transformer primary to avoid the possibility of accidental short circuits and shocks.

PARTS LIST

C_1, C_2—1.0μF, 250-volt, mylar film (Radio Shack 272-1055)
C_3—0.47-μF, 250-volt, mylar film (Radio Shack 272-1054)
C_4—0.22-μF, 250-volt mylar film (Radio Shack 272-1058)
C_5—0.1-μF, 250-volt, mylar film (Radio Shack 272-1053)
C_6—0.047-μF, 250-volt, mylar film (Radio Shack 272-1052)
S_1 through S_6—ON/OFF switches (Radio Shack 275-406)

Circuit 2-4. How to Stay Warm at Night and Still Reduce Your Fuel Bill

There's no need to turn down the thermostat of your central heating system to an uncomfortably low temperature at night to reduce fuel consumption. This circuit can allow you to remain comfortably warm while reducing your fuel bills even further. When plugged into a standard 24-hour household timer set to cycle-ON during your normal sleeping hours, it will disconnect your furnace from the living room-mounted thermostat, and connect it to one mounted in your master bedroom. Just close the heating ducts to the rest of your home almost OFF as you retire, and open those to your sleeping quarters (the opposite as you arise) for comfort plus increased savings.

For construction and installation, you can build this circuit into a decorative enclosure, and mount it in conjunction with the household timer alongside your living room thermostat. Install an additional thermostat of the same type used in the living room in your master bedroom. Finishing up, interconnect the thermostats to the DPDT relay as shown in the accompanying schematic diagram. Check your local house wiring codes (if any) before beginning this project.

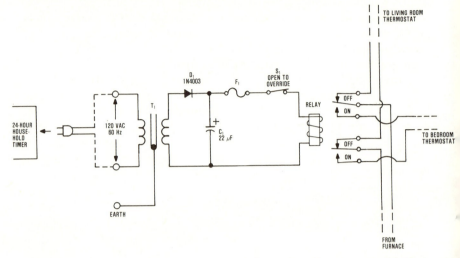

Circuit 2-4. How to stay warm at night and still reduce your fuel bill

PARTS LIST

C_1—22-μF, 25-volt aluminum electrolytic (Radio Shack 272-1014 or 272-1026)
D_1—IN4003 diode (Radio Shack 276-1102)
F_1—fuse to match current rating of transformer secondary
Relay—12-volt, DPDT miniature relay, 250 ohms or more coil resistance (Radio Shack 275-206)
S_1—ON/OFF switch (Radio Shack 275-602)
T_1—Miniature transformer with 12-volt or 12.6 volt secondary of at least double the relay's current demand (Radio Shack 273-1385)

Circuit 2-5. Temperature-Sensing Fan Regulator

Another circuit which provides economy plus comfort and convenience in your home or work environment. It automatically switches ON an electric fan as the ambient temperature exceeds any limit you select ... then OFF again as the temperature falls below this limit. This is great for use to switch ON your attic fan to vent hot, trapped air which causes your home or shop to be warmer than it has to be during summer days, then OFF in the

Semiconductor Circuits to Reduce Home Energy Use

evening as the temperature drops in order to save energy. Other practical applications include its use to control a stove exhaust fan, a soda fountain-type overhead fan, or a greenhouse ventilation fan.

For construction, use 12-volt power supply number 11-4 or

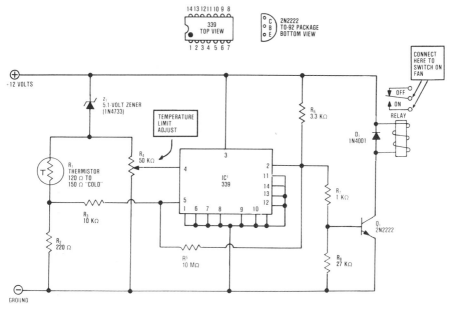

Circuit 2-5. Temperature-sensing fan regulator

PARTS LIST

D_1—IN4001 diode (Radio Shack 276-1101)
IC_1—339 integrated circuit (Radio Shack 276-1712)
Q_1—2N2222 or similar transistor (Radio Shack 276-1617)
R_1—120 ohm to 150 ohm "cold" resistance thermistor
R_2—220 ohms, ½-watt (Radio Shack 271-015)
R_3—10,000 ohm, ¼-watt (Radio Shack 271-1335)
R_4—50,000 ohm, ¼-watt (or more), linear taper potentiometer
R_5—10,000,000 ohm, ¼-watt (Radio Shack 271-1365)
R_6—3,300 ohm, ¼-watt (Radio Shack 271-1328)
R_7—1,000 ohm, ¼-watt (Radio Shack 271-1321)
R_8—27,000 ohm, ¼-watt (Radio Shack 271-1340)
Relay—12-volt, SPDT miniature relay, 250 ohms or more coil resistance (such as Radio Shack 275-206)
Z_1—5.1-volt zener diode (IN4733)

11-12 (see Chapter 11*) Mount "Temperature Limit Adjust" potentiometer R_4 for access from the front of the circuit box. Include a pointer-type knob on the shaft of this potentiometer to facilitate setting. Install thermistor R_1 remotely from the circuit in a location where it will accurately sense ambient room temperature. An electrical outlet built into the circuit box is usually the best way to connect with the fan. Be sure to select a relay with contacts of the appropriate current rating for your intended application.

For the initial circuit setting, there are several ways to go about it. One way is to simply wait until ambient temperature is where you want the circuit to activate, and then rotate Temperature Limit Adjust R_4 until the circuit just turns ON the fan. Another is to test this circuit during various times of the day in this same way, marking the temperature at which the circuit activates on the circuit box opposite the pointer knob with each trial for later reference.

Circuit 2-6. Pushbutton Switch That "Remembers"

Press switch S_1 to turn ON an appliance, and this circuit will "remember" to turn it OFF after a pre-selected interval—from about 30 seconds through 6 minutes. A few of the applications around your home where this circuit will not only reduce your power bill, but possibly your heating and air conditioning bills as well, include use with your kitchen or bathroom fan, heat lamp, blender, and garage light. As a few added bonuses, this circuit is compact enough to be built into an electrical wall box, and consumes no power at all until switch S_1 is pressed.

For construction, remember that high voltage will be present throughout much of this circuit and to act accordingly. Mount "Press To Start" switch S_1 and "ON Time Adjust" variable resistor R_3 for access from the front of the circuit enclosure. Notice that a DPDT relay is required, with one set of contacts used to latch-on this circuit once S_1 is pressed, the other to switch ON the appliance for the timed interval. Also, note that once this circuit is energized and then goes OFF, it must be allowed to "rest" for about a second before it can be energized again. Be sure to select a relay which has contact ratings adequate for the appliance to which it will be connected.

---

*For all references to power supply numbers (11-1, 11-4, 11-8, 11-12, etc.), see the circuits in Chapter 11.

Semiconductor Circuits to Reduce Home Energy Use 59

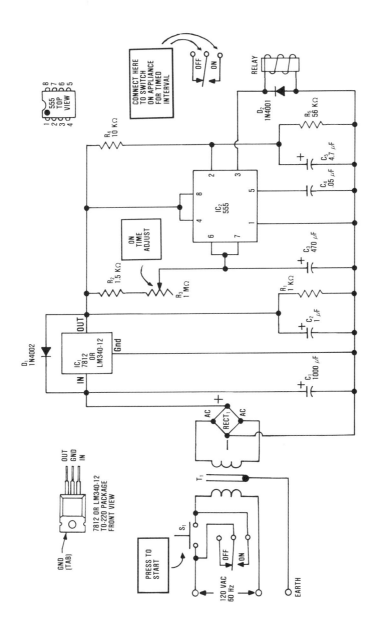

Circuit 2-6. Pushbutton switch that "remembers"

PARTS LIST

C_1—1,000-μF, 25-volt aluminum electrolytic (Radio Shack 272-1019 or 272-1032)
C_2—1-μF, 25-volt solid tantalum
C_3—470-μF, 25-volt aluminum electrolytic (Radio Shack 272-1018 or 272-1030)
C_4—0.05-μF, 50-volt ceramic disk (Radio Shack 272-134)
C_5—4.7-μF, 25-volt aluminum electrolytic (Radio Shack 272-1012 or 272-1024)
D_1—IN4002 diode (Radio Shack 276-1102)
D_2—IN4001 diode (Radio Shack 276-1101)
IC_1—7812 or LM340-12 3-terminal voltage regulator (Radio Shack 276-1771)
IC_2—555 integrated circuit (Radio Shack 276-1723)
R_1—1,000 ohms, ¼-watt (Radio Shack 271-1321)
R_2—1,500 ohms, ¼-watt (Radio Shack 271-025)
R_3—1,000,000 ohms, ¼-watt (or more), linear taper potentiometer (Radio Shack 271-211)
R_4—10,000 ohms, ¼-watt (Radio Shack 271-1335)
R_5—56,000 ohms, ¼-watt (Radio Shack 271-043)
$Rect_1$—1-amp, 50-PIV modular bridge rectifier (Radio Shack 276-1161)
Relay—12-volt, DPDT, miniature relay, 250 ohms or more coil resistance (Radio Shack 275-206)
S_1—Normally open push-button switch (Radio Shack 275-1547)
T_1—Miniature transformer with 12-volt, 250-mA to 500-mA secondary (Radio Shack 273-1385)

Circuit 2-7. Ambient Light Controller

Simply connect this circuit to your room lights in the order that you would normally turn them ON ... Set "Light Control" variable resistor R_3 for a comfortable level of illumination ... and this circuit will sequentially switch these lights ON and OFF to maintain this level henceforth. It's great for use as a light control for a work area which is partially illuminated by window lighting, such as your shop or kitchen, to slash energy use during the brighter times of the day. Works well with fluorescent strip lighting.

For construction, first remember that high voltage will be present throughout much of this circuit and take the necessary precautions. Use power supply number 11-5 and set it for 13 volts in order to provide extra power for closing the relays. Connect the first light that you would normally turn ON to the contacts of $Relay_1$, the second light to the contacts of $Relay_2$, etc. Omit the unneeded relays if you have less than four lights. Mount Light control R_3 and Trigger Level Adjust R_7 for access from the front of the circuit box. Be sure to install photoresistor R_2 where it will

Semiconductor Circuits to Reduce Home Energy Use 61

receive a representative sample of ambient room light and not be disturbed by shadows as people walk by. Select relays with contact ratings adequate for the lights to which they will be connected.

For the initial circuit adjustment, rotate Trigger Level Adjust R_7 for approximately 5,000 ohm resistance. Then, with the circuit energized, rotate Light Control R_3 to sequentially switch ON the number of needed lights—from one to four. If necessary, rotate Trigger Level Adjust R_7 for somewhat more resistance to increase circuit sensitivity, for somewhat less resistance to lower the threshold level when the first light comes ON.

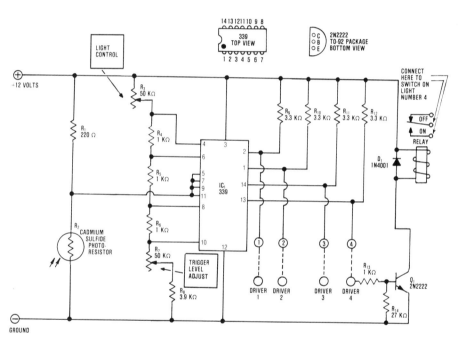

Circuit 2-7. Ambient light controller

PARTS LIST

D_1—IN4001 diode (Radio Shack 276-1101)
IC_1—339 integrated circuit (Radio Shack 276-1712)
Q_1—2N2222 or similar transistor (Radio Shack 276-1617)
R_1—220 ohm, ½-watt (Radio Shack 271-015)
R_2—Cadmium sulfide photoresistor, approximately 100 ohms light, 500,000 ohms dark (Radio Shack 276-116)
R_3, R_7—50,000 ohms, ¼-watt (or more), linear taper potentiometers (Radio Shack 271-1716)

R_4, R_5, R_6, R_{13}—1,000 ohm, ¼-watt (Radio Shack 271-1321)
R_8—3,900 ohm, ¼-watt (Radio Shack 271-1329)
R_9, R_{10}, R_{11}, R_{12}—3,300 ohm, ¼-watt (Radio Shack 271-1328)
R_{14}—27,000 ohm, ¼-watt (Radio Shack 271-1340)
Relay—12-volt, SPDT miniature relay, 250 ohms or more coil resistance (such as Radio Shack 275-206)

A few final notes: Although cost-per-unit of energy will likely go up, your total energy costs can remain the same or even decrease through more efficient use. Less can be more ... and semiconductors can help make this happen.

Using Semiconductors To Build Home Convenience Circuits and Projects

Semiconductors can serve a far wider variety of uses around your home than simply in television and audio units. This chapter will show you how to build circuits which can materially contribute to the level of safety and convenience around your home. Some of these circuits can even act as silent, hidden servants by performing simple tasks. And the energy-efficient nature and low cost characteristic of semiconductors make these units all the more practical.

One circuit which can add to your piece of mind is Battle Lantern Circuit 3-1. Your house power can fail for numerous reasons including a downed line, a short circuit, or a fire. But whatever the reason for the failure, it's comforting to know that this circuit will automatically switch ON a battery-powered lantern to illuminate some critical area of your home such as an escape route or a fuse box.

On the theme of convenience, a circuit which can benefit most households is the Doorbell of Two Tones (Circuit 3-2). Rather than having you guess whether your caller is at the front door or at the back door, this circuit lets you know by sounding a different tone as respective door buttons are pressed. Another beneficial electronic mechanism is the Rain Alarm (Circuit 3-3). It will keep a close watch for rain (helpful when you have wash on the line), and give you advance warning as the first few drops fall.

Although accepting the "thorns with the roses" is a fact of life in most instances, the T.V. Commercial Killer shown in Circuit 3-4 makes such inconvenience unnecessary when watching television programs. Simply click it to the "Kill Sound" position when an annoying commercial comes on and it will do away with the audio portion for an adjustable time interval. Then when this interval is up, it automatically restores sound for the continuation of the program.

As a few more examples on how to live better through high technology, the Electronic Scarecrow of Circuit 3-7 can discourage birds, squirrels, deer and other freeloaders from raiding your vegetable patch or fruit trees while produce is ripening. It does so by sounding a threatening tone at periodical interval—a tone which can even be adjusted for ultra-sonic operation in order to keep the neighbors happy. And the Electric Flypaper of Circuit 3-8 can rid your home of pesty flying insects without the need for toxic (and expensive) sprays.

Circuit 3-1. Battle Lantern

Never trust a line-powered light to work during an emergency. House fires are often caused by short circuits in the wiring which disrupt power, and floods and hurricanes invariably take down poles. Lights often fail just when you need them most.

Here's a highly reliable circuit which can show you the way out of a threatened building, or illuminate an important piece of gear such as a control panel or fuse box. When house power fails, it

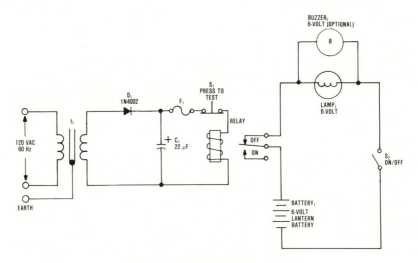

Circuit 3-1. Battle lantern

automatically switches ON a strategically placed battery-powered lantern. An optional buzzer sounds an audio tone for helping occupants find their way out of a smoke-filled building when visibility is obscured.

For construction, almost any 12-volt SPDT relay will do to switch ON the emergency lamp (and buzzer). For maximum long-term reliability, be sure to use a 12-volt transformer with a current rating at least double that required by the relay. The battery-powered light may be a hand-lantern which is wired as shown in the accompanying schematic (Circuit 3-1) and carefully aimed for best effect ... or constructed from separate parts in a decorative yet functional enclosure.

Be sure to test the unit at least monthly. To do this, push "Press To Test" switch S_1 while the unit is energized and watch for the lamp to glow brightly (and listen for the buzzer to sound).

Using Semiconductors for Home Conveniences 67

PARTS LIST

Battery$_1$—6-volt lantern-type battery of sufficient current capability to drive lamp (and buzzer) (Radio Shack 23-066)
Buzzer$_1$ (optional)—6-volt buzzer (Radio Shack 273-060 or 273-049)
C$_1$—22-μF, 25-volt aluminum electrolytic (Radio Shack 272-1014 or 272-1026)
D$_1$—IN4002 diode (Radio Shack 276-1102)
F$_1$—Fuse to match current rating of transformer secondary
Lamp$_1$—6-volt lamp (such as Radio Shack 272-1142)
Relay$_1$—12-volt SPDT miniature relay (Radio Shack 275-206)
S$_1$—Normally closed pushbutton switch (Radio Shack 275-1548)
S$_2$—ON/OFF switch (Radio Shack 275-602)
T$_1$—Miniature transformer with 12-volt or 12.6-volt secondary of at least double the relay's current demand (Radio Shack 273-1385)

Circuit 3-2. Doorbell of Two Tones

Never again need you mistakenly answer the front door when your caller is really at the back door. This electronic doorbell distinguishes between doors by sounding a warble tone when door button S$_1$ is pressed ... a steady tone when door button S$_2$ is pressed. And it is practical for use with a battery power supply as it consumes no energy at all until a door button is pressed.

For construction, you may use line-powered supply 11-4, 11-5, 11-11, or 11-12, depending on the current demands of your bell or

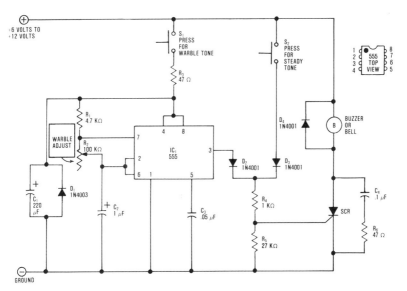

Circuit 3-2. Doorbell of two tones

buzzer. Or else you may use a 6-volt or a 12-volt lantern-type battery to simplify construction. Note that a self-interrupting type bell or buzzer must be used so that the sounder will disconnect when the door button is released. Almost all sounders are of this type, but be sure to check for proper operation before making a permanent installation. Position "Warble Adjust" variable resistor R_2 for easy access from the front of the circuit box. Interconnect door buttons S_1 and S_2 to the circuit using standard 2-conductor doorbell wire.

For the initial circuit adjustment, set Warble Adjust R_2 to its mid-position. Then have someone press-and-hold door button S_1 while you rotate R_2 until the desired tone is obtained.

PARTS LIST

Buzzer or Bell—6-volt to 12-volt (to match power supply), self-interrupting type buzzer or bell
C_1—220-μF, 25-volt aluminum electrolytic (Radio Shack 272-1017 or 272-1029)
C_2—1-μF, 25-volt aluminum electrolytic (Radio Shack 272-1419)
C_3—0.05-μF, 50-volt ceramic disk (Radio Shack 272-134)
C_4—0.1-μF, 50-volt ceramic disk (Radio Shack 272-135)
D_1—IN4003 diode (Radio Shack 276-1102)
D_2, D_3, D_4—IN4001 diodes (Radio Shack 276-1101)
IC_1—555 integrated circuit (Radio Shack 276-1723)
R_1—4,700 ohm, ¼-watt (Radio Shack 271-1330)
R_2—100,000 ohm, ¼-watt (or more) linear taper potentiometer (Radio Shack 271-092)
R_3, R_6—47 ohm, ¼-watt (Radio Shack 271-1307)
R_4—1,000 ohm, ¼-watt (Radio Shack 271-1321)
R_5—27,000 ohm, ¼-watt (Radio Shack 271-1340)
S_1, S_2—Normally open pushbutton switches
SCR—Any medium-current silicon controlled rectifier (such as Radio Shack 276-1067)

Circuit 3-3. Rain Alarm

For early warning of rain so that you can remove clothes from the line and close your windows. When even the slightest splatter of rain bridges the elements of the contact grid, this circuit sounds a continuous tone and lights an LED to give you advanced notice. It can be easily reset when the rain has ceased by blowing the contact grid free of moisture and re-energizing the circuit.

Using Semiconductors for Home Conveniences 69

For construction, you can either use line-powered supply 11-4, 11-11, or 11-12 ... or battery supply 11-18 to simplify circuit construction. The contact grid may be as simple as two bare wires positioned very close, but not touching. Position contact grid where splatters will quickly bridge the air gap between adjacent elements as the rain commences. The "Sensitivity Adjust" variable resistor R_2 may be mounted on the circuit board as it rarely needs adjustment beyond the initial setting. Initially, set this control to its mid-position.

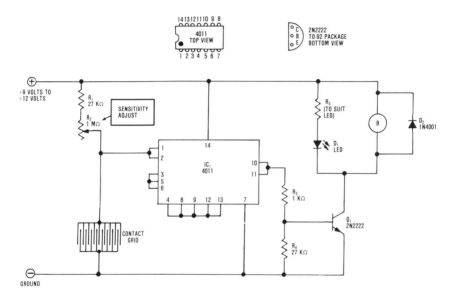

Circuit 3-3. Rain alarm

PARTS LIST

Buzzer—9-volt to 12-volt piezo buzzer (Radio Shack 273-060)
D_1—Any red light-emitting diode (such as Radio Shack 276-026)
D_2—1N4001 diode (Radio Shack 276-1101)
IC_1—4011 integrated circuit (Radio Shack 276-2411)
Q_1—2N2222 or similar transistor (Radio Shack 276-1617)
R_1, R_4—27,000 ohm, ¼-watt (Radio Shack 271-1340)
R_2—1,000,000 ohm, ¼-watt (or more) linear taper potentiometer (Radio Shack 271-229)
R_3—1,000 ohm, ¼-watt (Radio Shack 271-1321)
R_5—To suit LED

Circuit 3-4. T.V. Commercial Killer

A circuit which will eliminate the most unwanted portion of your television (or radio) programs—the sound of commercials. Simply click 2-position switch S_1 to the "Kill Sound" position, and it will do away with the unwanted audio for a timed period that is easily adjustable to suit the scheduling of any given program (from about 30 seconds through 4 minutes). Then it will automatically let you know when the program comes back on by

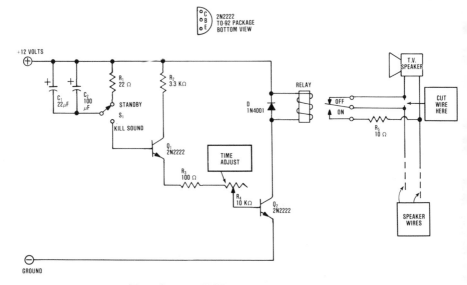

Circuit 3-4. T.V. commercial killer

restoring sound. Sound may be immediately restored at any time during this interval by clicking switch S_1 to the "Standby" position.

For construction, use plug-in supply number 11-4 (adjustable), or 11-12. If possible, use the adjustable supply and set it for 13 volts to provide extra power for closing the relay. Position "Time Adjust" variable resistor R_4 for easy front-panel access. Also, use a knob with a pointer so that you can inscribe calibration marks on the front of the circuit box to suit various-length commercials. The "Standby/Kill Sound" switch S_1 is best mounted in a hand-held unit to allow quick switching with minimal effort.

When removing the back of the T.V. set for making connection with the speaker, beware the high-voltage components inside which can hold a dangerous charge even though

Using Semiconductors for Home Conveniences 71

the line cord is removed from power. If you don't know where these components are located, have an experienced technician show you.

Note that switch S_1 must be placed in the Standby position for one-half second or longer before being placed in the Kill Sound position for best results.

PARTS LIST

C_1—22-μF, 25-volt aluminum electrolytic (Radio Shack 272-1014 or 272-1026)
C_2—100-μF, 25-volt aluminum electrolytic (Radio Shack 272-1016 or 272-1028)
D_1—IN4001 diode (Radio Shack 276-1101)
Q_1, Q_2—2N2222 or similar transistors (Radio Shack 276-1617)
R_1—22 ohms, ¼-watt (Radio Shack 271-005)
R_2—3,300 ohms, ¼-watt (Radio Shack 271-1328)
R_3—100 ohms, ¼-watt (Radio Shack 271-1311)
R_4—10,000 ohms, ¼-watt (or more), linear taper potentiometer (Radio Shack 271-1715)
R_5—10 ohm, 10-watt (Radio Shack 271-132)
Relay—12-volt SPDT relay, 250 ohms or more coil resistance (Radio Shack 275-206)
S_1—SPDT switch (Radio Shack 275-613)

Circuit 3-5. Carbon-Zinc Battery Refresher

Prolong the life of standard dry-cell batteries! Although carbon-zinc cells are technically not rechargeable, they may be successfully "refreshed" a dozen times or more to vastly extend their service life. Best results are obtained with fairly new, only partially discharged cells. For rejuvenation, this circuit outputs a gentle 50 mA of current into any single or series combination of cells as long as their terminal voltage does not exceed about 10 volts. And for automatic cut-off when the charging interval is over (see chart which accompanies schematic (Circuit 3-5) for recommended intervals) simply plug this circuit into a household timer. It may also be used for recharging most varieties of nickel-cadmium batteries (use charge times recommended for those batteries).

For construction and operating tips, install components inside a circuit box with color-coded plus-and-minus attachment leads for the cells to be charged. A series-wired battery holder is usually the most convenient way to mount cylindrical-shaped batteries for charging, and a standard clip for 9-volt, transistor radio-type batteries. Once a battery is charged, it should be allowed to "rest" for six to eight hours before being put back into service.

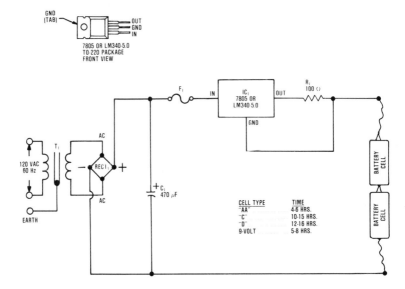

Circuit 3-5. Carbon-zinc battery refresher

PARTS LIST

C_1—470-μF, 25-volt aluminum electrolytic (Radio Shack 272-1018 or 272-1030)
F_1—Choose to match current rating of transformer secondary
IC_1—7805 or LM340-5.0, 3-terminal voltage regulator (Radio Shack 276-1770)
R_1—100 ohm, ½-watt (Radio Shack 271-012)
$Rect_1$—1-amp, 50 PIV, modular bridge rectifier (Radio Shack 276-1161)
T_1—Miniature transformer with 12-volt or 12.6-volt, up to 500-mA secondary (Radio Shack 273-1385)

Circuit 3-6. Automatic Pump Control Unit

Never again will you have to get up in the middle of the night to pump out your flooded basement during the rainy season, or rely on an often-troublesome mechanical control. As drainage in your collection sump accumulates to the level of Sensor Probe$_2$,

Using Semiconductors for Home Conveniences 73

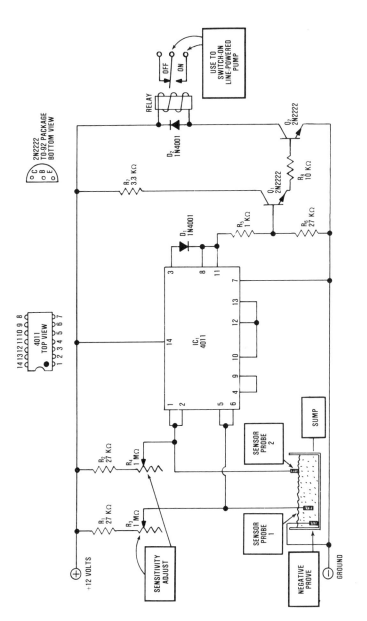

Circuit 3-6. Automatic pump-control unit

this solid state circuit automatically turns ON your line-powered pump ... then OFF again as the liquid level falls below Sensor Probe$_1$. And this circuit consumes so little power that you can leave it on all year to guard against such hazards as broken water pipes.

For construction, use 12-volt supply 11-4 (adjustable) or 11-12. If possible, use the adjustable supply and set it for 13 volts in order to provide extra power for closing the relay. Set perfboard mounted "Sensitivity Adjust" variable resistors R_3 and R_4 to their mid-positions (in only rare installations do they need further adjustment for more or less sensitivity). Locate the Negative Probe low in the sump ... Sensor Probe$_1$ and Sensor Probe$_2$ freely suspended directly above. Almost any metallic objects, such as bolts, will serve for probes. Be sure to install the circuit box in a dry, protected location to prevent corrosion. Choose a relay with contact ratings adequate for the pump which is to be attached.

PARTS LIST

D_1, D_2—IN4001 diodes (Radio Shack 276-1101)
IC_1—4011 integrated circuit (Radio Shack 276-2411)
Q_1, Q_2—2N2222 or similar transistors (Radio Shack 276-1617)
R_1, R_2, R_6—27,000, ohm, ¼-watt (Radio Shack 271-1340)
R_3, R_4—1,000,000 ohm, ¼-watt, linear taper potentiometers (Radio Shack 271-229)
R_5—1,000 ohm, ¼-watt (Radio Shack 271-1321)
R_7—3,300 ohm, ¼-watt (Radio Shack 271-1328)
R_8—10,000 ohm, ¼-watt (Radio Shack 271-1335)
Relay—12-volt, SPDT relay, 160 ohms or more coil resistance (such as Radio Shack 275-218)

Circuit 3-7. Electronic Scarecrow

Rid your vegetable garden and fruit orchard of birds, squirrels, deer and other pests the easy humane way. Simply position the speaker in the area to be protected and this circuit will sound a loud, raucous tone at timed intervals. The "Tone Synthesis" variable resistors R_5 and R_6 allow you to alter the tone periodically so that pests can't become complacent. You can even create a tone at a frequency above that of human hearing (but within the range of animal hearing) so as not to disturb the neighbors.

For construction, use 12-volt plug-in supply 11-4 or 11-12 and route the speaker to the area to be protected via a speaker wire. In cases where there is no electrical outlet in the vicinity,

Using Semiconductors for Home Conveniences 75

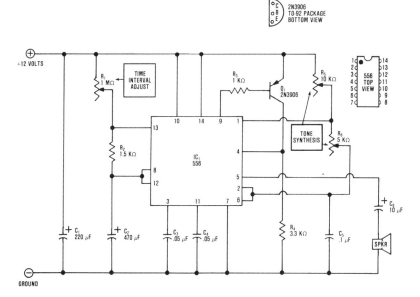

Circuit 3-7. Electronic scarecrow

PARTS LIST

C_1—220-μF, 25-volt aluminum electrolytic (Radio Shack 272-1017 or 272-1029)
C_2—470-μF, 25-volt aluminum electrolytic (Radio Shack 272-1018 or 272-1030)
C_3, C_4—0.05-μF, 50-volt ceramic disk (Radio Shack 272-134)
C_5—0.1-μF, 50-volt ceramic disk (Radio Shack 272-135)
C_6—10-μF, 25-volt aluminum electrolytic (Radio Shack 272-1013 or 272-1025)
IC_1—556 integrated circuit (Radio Shack 276-1728)
Q_1—2N3906 or similar transistor (Radio Shack 276-2034)
R_1—1,000,000 ohm, ¼-watt (or more), linear-taper potentiometer (Radio Shack 271-211)
R_2—1,500 ohm, ¼-watt (Radio Shack 271-025)
R_3—1,000 ohm, ¼-watt (Radio Shack 271-1321)
R_4—3,300 ohm, ¼-watt (Radio Shack 271-1328)
R_5—10,000 ohm, ¼-watt (or more), linear-taper potentiometer (Radio Shack 271-1715)
R_6—5,000 ohm, ¼-watt (or more), linear-taper potentiometer (Radio Shack 271-1714)
SPKR—Any 8-ohm speaker (use a "tweeter"-type for ultrasonic service)

---

battery supply 11-20 should power this circuit for an entire growing season on a single charging. Position Time Interval

76   *Using Semiconductors for Home Conveniences*

Adjust variable resistor R_1 plus Tone Synthesis variable resistors R_5 and R_6 for easy access from the front of the circuit box. Note that the speaker should be mounted in a weather-tight enclosure. For ultra-sonic service, a "tweeter" type speaker is recommended for best efficiency.

To adjust this circuit for operation, begin by rotating Time Interval Adjust R_1 for minimum resistance so that the circuit sounds in rapid succession. Next, adjust Tone Synthesis resistors R_5 and R_6 in combination for the desired tone. For ultra-sonic service, tone is best set with the aid of an oscilloscope. Finishing up, rotate R_1 for the desired time interval between tone bursts (up to about 6 minutes is possible).

Circuit 3-8. Electronic Flypaper

Rid your home or patio of unwanted flying insects without the use of toxic chemicals. Housed in a decorative enclosure, this unit hangs from the ceiling or a lamppost out of the reach of children and pets. Then when a fly, mosquito, or other winged

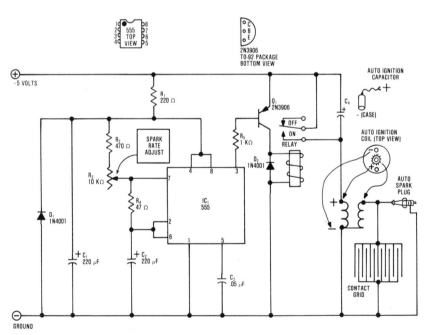

Circuit 3-8. Electronic flypaper

Using Semiconductors for Home Conveniences 77

insect is attracted to the contact grid by a "black light" inside the enclosure, it is painlessly zapped. If the contact grid is accidentally touched by a human or pet, only a harmless (though resounding) shock is received.

For construction, use 5-volt supply number 11-3. Position "Spark Rate Adjust" variable resistor R_3 for easy servicing. As a suggestion for making the contact grid, wrap spiraling turns of bare copper wire down a wooden framework with about 5 mm of space between each turn. Next, wrap a second layer of turns, interleaving the first with about 2½ mm of space between adjacent turns. Finally, connect each wire to the circuit as shown using spark-plug type wire.

The spark plug serves to dissipate energy not consumed by unwary insects. Use a new plug, and set the gap between electrodes so that the spark arcs across here rather than the contact grid elements. Because this spark may serve as a source of unwanted ignition, keep inflammable liquid/vapors a safe distance away.

PARTS LIST

Auto ignition capacitor—Any 12-volt auto ignition capacitor
Auto ignition coil—Any 12-volt auto ignition coil
Auto spark plug—Any auto spark plug
C_1, C_2—220-μF, 25-volt aluminum electrolytic (Radio Shack 272-1017 or 272-1029)
C_3—0.05-μF, 50-volt ceramic disk (Radio Shack 272-134)
D_1, D_2—1N4001 diodes (Radio Shack 276-1101)
IC_1—555 integrated circuit (Radio Shack 276-1723)
Q_1—2N3906 or similar transistor (Radio Shack 276-2034)
R_1—220 ohm, ¼-watt (Radio Shack 272-1313)
R_2—470 ohm, ¼-watt (Radio Shack 271-1317)
R_3—10,000 ohm, ¼-watt (or more), linear-taper potentiometer (Radio Shack 271-1715)
R_4—47 ohm, ¼-watt (Radio Shack 271-1307)
R_5—1,000 ohm, ¼-watt (Radio Shack 271-1321)
Relay—SPDT 5-volt relay, 50 ohms or more coil resistance (Radio Shack 275-215)

Summing up, electronics can provide pragmatic solutions to problems which previously had to be dealt with through mechanical or chemical means—if at all. This is just one more instance where semiconductor technology can bring convenience—and new opportunities for profit—to you.

Circuits for Harvesting the Free Energy of the Sun

There's every good reason for you to take advantage of solar energy in the 80's—especially if you live in a locality which receives fairly strong light for the greater portion of the year. Solar energy is already fairly concentrated ... for all practical purposes inexhaustible ... and above all absolutely free. The instructions in this chapter will show you how to build innovative circuits to harvest the energy of the sun.

Probably the most simplistic way to collect the sun's energy is with the Circuit 4-1 Solar Electric Generator. Here, individual solar cells soak up this radiation and convert it directly into dc electricity for any use you choose. One decided advantage to this circuit is that solar cells last practically forever as long as they are not subjected to severe physical stress.

Circuit 4-2 shows you how to put a solar electric generator to immediate use in a practical application. The Self-Powered Strobe Light/Beacon Flasher described is as utilitarian as it is amazing!

To date, however, the most cost-effective way to harvest solar energy is by gathering its heat. The Circuit 4-4 Solar Energy Trap can show you one direct way to do this. It will open a curtain or skylight in the early morning to admit healthful heat and light into your home during the day, and then shut the curtain/skylight in the evening to trap in accumulated heat.

Presently, solar collectors are rapidly gaining popularity as an effective means of heating water for use in home, pool, and elsewhere. For regulating the flow of water for maximum heat absorption, the Circuit 4-3 pump control will constantly monitor the temperature in both your solar array and collection tank (or pool), and then switch a circulating pump ON and OFF as appropriate.

As another example of a circuit for use with a solar collector, Left-to-Right Sun Tracker Circuit 4-5 can increase the sunlight-gathering efficiency of your array by a factor of two or more by keeping it pointed directly at the sun throughout the day. And when used in conjunction with other circuits found in this chapter, it can provide completely automatic operation at a small fraction of the cost for an equivalent commercial unit.

Circuit 4-1. Solar Electric Generator

Converting sunlight directly into electric energy via solar cells is already practical in many applications, and will become more so as the price of these devices goes down and their efficiency goes up. Worthwhile applications include use in building self-charging bicycle and camping lights, electronic circuits, and for trickle-charging automotive-type batteries for improved service and longer life.

In this simple though eminently practical solar generator, individual solar cells with the standard output voltage of about 0.45 V each are wired in series. Cells are available with current capabilities from about 25 mA through 1 amp. Choose cells to suit your application. Their combined energy can directly power a load during the brighter times of the day, with excess energy being absorbed by the storage batteries for use when ambient light is low. Diode D_1 prevents battery current from flowing through the solar cells when their output voltage is below that of the batteries.

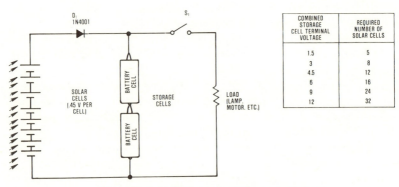

Circuit 4-1. Solar electric generator

COMBINED STORAGE CELL TERMINAL VOLTAGE	REQUIRED NUMBER OF SOLAR CELLS
1.5	5
3	8
4.5	12
6	16
9	24
12	32

PARTS LIST

D_1—1N4001 diode (Radio Shack 276-1101)
Load—Any low-current load such as a small incandescent bulb, motor winding, etc.
S_1—ON/OFF switch (Radio Shack 275-602)
Solar Cells—Silicon solar cells connected in series (size of necessary solar cells depends on such factors as daily current draw, season, and weather).
Storage Cells—Any battery cells capable of taking a charge. nickel-cadmium and lead-acid cells work well, although standard carbon-zinc cells are generally satisfactory. Use larger cells to provide more storage capability.

Circuits for Harvesting the Free Energy of the Sun

The storage cells may be almost any type capable of taking a charge. Nickel-cadmium and lead-acid cells work well, although standard carbon-zinc cells are generally satisfactory. The load may be almost any incandescent bulb, electronic circuit, or motor with a total energy draw less than or equal to that produced by the solar cells on an average day.

Circuit 4-2. Self-Powered Strobe Light/Beacon Flasher

A self-powered circuit which periodically emits a bright pulse of light for the purpose of gaining attention at night. Visibility can actually be over ten miles in clear weather! When adjusted for a rapidly re-occurring pulse, this circuit is excellent for use as an anti-collision light for a bicycle, boat or pedestrian. Or when adjusted for a slowly re-occurring pulse, it can be left on continuously as a beacon flasher to mark a boat mooring float, pier landing, or other marine hazard.

To assemble and adjust this circuit, first choose a circuit enclosure to suit the intended application. In most cases this means mounting circuit components in a weather-tight box with provisions so that the solar cells will receive full sunlight during

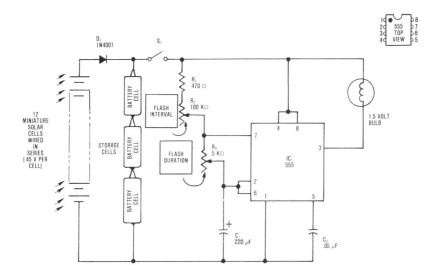

Circuit 4-2. Self-powered strobe light/beacon flasher

the day. Position the lamp bulb for optimum visibility. Next, rotate "Flash Interval" variable resistor R_2 to adjust the time interval between pulses to suit your particular application ... and then "Flash Duration" variable resistor R_3 for a quick, bright pulse. Remember that you may have to go back and forth between these controls several times because there is some degree of interaction.

PARTS LIST

C_1—220-μF, 25-volt aluminum electrolytic (Radio Shack 272-1017 or 272-1029)
C_2—0.05-μF, 50-volt ceramic disk (Radio Shack 272-134)
D_1—1N4001 diode (Radio Shack 276-1101)
IC_1—555 integrated circuit (Radio Shack 276-1723)
R_1—470 ohm, ¼-watt (Radio Shack 271-1317)
R_2—100,000 ohms, ¼-watt or more, linear taper potentiometer (Radio Shack 271-092)
R_3—5,000 ohms, ¼-watt or more, linear taper potentiometer (Radio Shack 271-1714)
Storage Cells—Either three carbon-zinc cells, or four nickel-cadmium cells. Choose size to match energy storage requirements (varies according to climate)
S_1—ON/OFF switch (Radio Shack 275-602)
1.5 volt bulb—1.5-volt miniature incandescent bulb, 15 to 25 mA.
12 minature solar cells—12 miniature solar cells wired in series, each of 35 mA or higher current rating

Circuit 4-3. Pump Control for Solar Water Heater

An automatic control which continuously monitors the water temperature in both your solar collector and storage tank (or swimming pool), and switches ON a circulation pump when the temperature in the collector exceeds that in the tank by an adjustable amount. And because it senses the relative temperature of these two containers (rather than absolute temperature), it can function year-round in many localities for maximum benefit. As an added advantage, once it switches the pump ON, it holds it ON for a minimum of 30 seconds in order to prevent "hunting."

For construction, use 12-volt supply number 11-4 or 11-12 (Chapter 11). Position "Temperature Offset Adjust" variable resitor R_2 for easy access from the front of the circuit box.

Circuits for Harvesting the Free Energy of the Sun 85

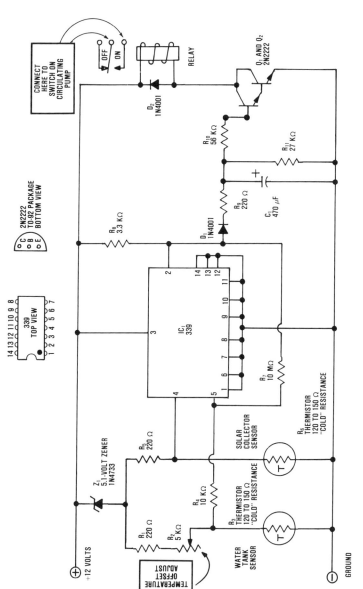

Circuit 4-3. Pump control for solar water heater.

Thermistors R_3 and R_6, which are routed between circuit box and tanks via 2-conductor speaker-type wire, must be fully emerged in their respective water containers and out of contact with the sidewalls to accurately read ambient temperatures. Be sure to use a relay with contact ratings adequate for the pump to which it will be connected.

To initially adjust this circuit, first turn Temperature Offset Adjust R_2 for zero ohms ... then gradually increase resistance for best circuit performance through a trial-and-error procedure. The optimum setting of this control will differ from application to application.

PARTS LIST

C_1—470-μF, 25-volt aluminum electrolytic (Radio Shack 272-1018 or 272-1030)
D_1, D_2—IN4001 diodes (Radio Shack 276-1101)
IC_1—339 integrated circuit (Radio Shack 276-1712)
Q_1, Q_2—2N2222 or similar transistors (Radio Shack 276-1617)
R_1, R_5, R_9—220 ohm, ¼-watt (Radio Shack 271-1313)
R_2—5,000 ohm, ¼-watt or more, linear taper potentiometer (Radio Shack 271-1714)
R_3, R_6—120 ohm to 150 ohm "cold" resistance thermistor
R_4—10,000 ohm, ¼-watt (Radio Shack 271-1335)
R_7—10,000,000 ohm, ¼-watt (Radio Shack 271-1365)
R_8—3,300 ohm, ¼-watt (Radio Shack 271-1328)
R_{10}—56,000 ohm, ¼-watt (Radio Shack 271-043)
R_{11}—27,000 ohm, ¼-watt (Radio Shack 271-1340)
Relay—12-volt, SPDT relay, 160 ohm or more coil resistance (such as Radio Shack 275-218)
Z_1—5.1 volt zener diode (1N4733)

Circuit 4-4. Solar Energy Trap

Acts as a one-way energy valve to admit healthful solar heat and light into your home during the day, and then to seal in warmth during the night. As light strikes photoresistor R_2 shortly after dawn, this circuit triggers to open the drapes from a south-facing window or skylight via a small motor—the opposite as light begins to fail at dusk. Not only will it reduce your energy bill, but it will make your home a brighter and more cheerful place to live in.

For construction, use 12-volt power supply number 11-5. Mount components inside a circuit box, with front panel access for

Circuits for Harvesting the Free Energy of the Sun 87

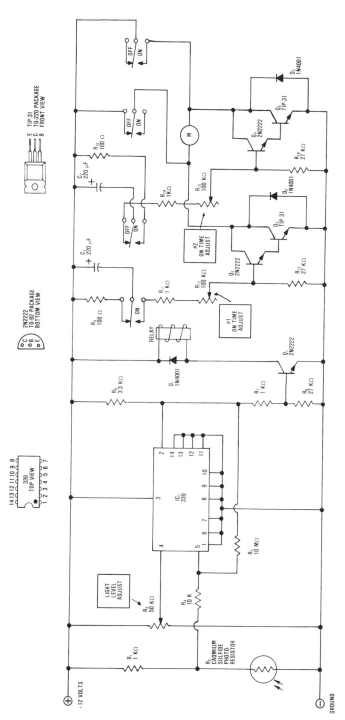

Circuit 4-4. Solar energy trap

PARTS LIST

C_1, C_2—220-μF, 25-volt aluminum electrolytic (Radio Shack 272-1017 or 272-1029)
D_1, D_2, D_3—1N4001 diodes (Radio Shack 276-1101)
IC_1—339 integrated circuit (Radio Shack 276-1712)
M—Any 12-volt, low-current reversible motor such as that from a scrap tape recorder
Q_1, Q_2, Q_4—2N2222 or similar transistors (Radio Shack 276-1617)
Q_3, Q_5—TIP 31 or similar transistor (Radio Shack 276-2017)
R_1, R_7, R_{11}, R_{14}—1,000 ohm, ¼-watt (Radio Shack 271-1321)
R_2—Cadmium sulfide photoresistor (about 500,000 ohms dark, 100 ohms light) (Radio Shack 276-116)
R_3—50,000 ohms, linear taper potentiometer, ¼-watt or more (Radio Shack 271-1716)
R_4—10,000 ohms, ¼-watt (Radio Shack 271-1335)
R_5—10,000,000 ohms, ¼-watt (Radio Shack 271-1365)
R_6—3,300 ohms, ¼-watt (Radio Shack 271-1328)
R_8, R_{13}, R_{16}—27,000 ohms, ¼-watt (Radio Shack 271-1340)
R_9, R_{10}—100 ohms, ¼-watt (Radio Shack 271-1311)
R_{12}, R_{15}—100,000 ohms, ¼-watt or more, linear taper potentiometers (Radio Shack 271-092)
Relay—12-volt, 4PDT relay, 160 ohms or more coil resistance (Radio Shack 275-214)

"Light Level Adjust" potentiometer R_3, "#1 ON Time Adjust" variable resistor R_{12}, and "#2 ON Time Adjust" variable resistor R_{15}. Photoresistor R_2 may be located either inside the circuit box with a hole to admit light, or externally via 2-conductor speaker wire. Be sure, however, that it is positioned where it will receive a representative amount of light and not be falsely triggered by automobile headlamps, etc.

The mechanical arrangement to open the drapes may be as simple as a revolving drum driven by the small, reversible motor. In the morning, the circuitry will activate the geared-down motor to turn the drum one way, winching the drapes open. In the evening, the circuitry will activate the motor to turn the drum the other way, winching the drapes shut. A belt drive arrangement which allows slippage is effective as it makes the motor ON time adjustment less critical.

For the initial circuit adjustment, close the drapes and energize the circuitry. With an amount of light striking photoresistor R_2 to simulate dawn, rotate Light Level Adjust R_3 until the circuit just activates the motor (if the motor turns the wrong way, simply reverse the motor's connections). Next, turn #1 ON Time Adjust R_{12} to adjust for sufficient motor ON time to completely open the drape. Finishing up, place your finger over

Circuits for Harvesting the Free Energy of the Sun **89**

the face of photoresistor R_1 to simulate dusk, and turn #2 ON Time Adjust R_{15} to adjust for sufficient motor ON time to completely shut the drapes. If for some reason you need more motor ON time, substitute larger-value capacitors for C_1 and C_2.

Circuit 4-5. Left-to-Right Sun Tracker for Your Solar Array

This solid-state control will greatly increase the energy-gathering ability of your solar array by rotating it on a hinge throughout the day to keep it pointed at the sun. As the sun moves across the sky, light passing through a narrow slot in a small enclosure strikes photoresistor R_1. In turn, this causes the circuit to latch ON a small motor for a precisely adjustable time interval thereby incrementing the array a few degrees in the direction of the sun's movement via a mechanical arrangement. This cycle then repeats throughout the solar day.

At dusk, you must either manually rotate the array back into position for pickup of light at dawn, or else use Return-To-Dawn Circuit 4-6 for automatic operation. Also, for best performance, you must periodically adjust the array up/down to compensate for seasonal changes in the sun's altitude. For automatic operation hour-to-hour, for such applications as accurately focusing light on a silicon solar panel, you can use Up/Down Control Circuit 4-7.

For construction, use either 12-volt plug-in supply number 11-5 or Solar Electric-Generator Circuit 4-1 in order to make this unit self-powered. Position "Light Level Adjust" potentiometer R_3 and "ON Time Adjust" variable resistor R_9 for access from the front of the circuit box. Install the R_1 photoresistor inside a small enclosure which has a light-absorbent inner coating plus a vertical slot to admit light, positioning this photoresistor within the enclosure for best effect with a trial-and-error procedure. Next, connect the motor to the array with a geared-down arrangement so that as the motor spins, it slowly rotates the array.

To adjust this circuit, shine a flashlight through the vertical slot in the small enclosure box onto the photoresistor, and then turn Light Level Adjust R_3 until the circuit just clicks the motor ON. If the motor rotates the array in the wrong direction, simply reverse the motor's connections. Next, turn ON Time Adjust R_9 for sufficient motor ON time to increment the array a few degrees in the direction of the sun.

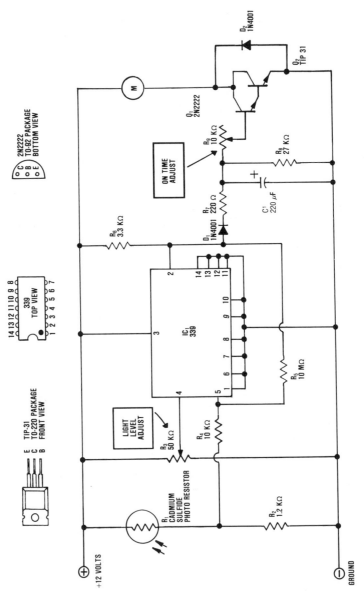

Circuit 4-5. Left-to-right sun tracker for your own solar array

Circuits for Harvesting the Free Energy of the Sun 91

PARTS LIST

C_1—220-μF, 25-volt aluminum electrolytic (Radio Shack 272-1017 or 272-1029)
D_1, D_2—1N4001 diodes (Radio Shack 276-1101)
IC_1—339 integrated circuit (Radio Shack 276-1712)
M—Any 12-volt, low-current, reversible motor such as that from a scrap tape recorder
Q_1—2N2222 or similar transistor (Radio Shack 276-1617)
Q_2—TIP 31 or similar transistor (Radio Shack 276-2017)
R_1—Cadmium sulfide photoresistor, about 500,000 ohms, dark, 100 ohms light (Radio Shack 276-116)
R_2—1,200 ohms, ¼-watt (Radio Shack 271-024)
R_3—50,000 ohms, ¼-watt or more, linear taper potentiometer (Radio Shack 271-1716)
R_4—10,000 ohms, ¼-watt (Radio Shack 271-1335)
R_5—10,000,000 ohms, ¼-watt (Radio Shack 271-1365)
R_6—3,300 ohms, ¼-watt (Radio Shack 271-1328)
R_7—220 ohms, ¼-watt (Radio Shack 271-1313)
R_8—27,000 ohms, ¼-watt (Radio Shack 271-1340)
R_9—10,000 ohms, ¼-watt or more, linear taper potentiometer (Radio Shack 271-1715)

Circuit 4-6. Return-to-Dawn Control for Your Solar Array

Here is a circuit that will automatically rotate your hinged solar array back to its dawn position at a preselected time each evening. You can use it in conjunction with almost any left-to-right sun tracking mechanism, or connect it directly into Circuit 4-5 with the diagram pictured in Figure 4-1.

In operation, as the rotating solar array mechanically contacts "Limit Switch 2" (which is positioned for contact at dusk), it closes this switch thereby latching-ON the relay. The relay activates a small motor, which in turn rotates the solar array back toward its dawn position via a mechanical geared-down drive arrangement. Then as the array mechanically contacts "Limit Switch 1" (which is positioned for contact at the dawn position), it opens this switch thereby de-activating the relay and stopping the motor.

For construction, use power supply 11-5. If you are using this circuit in conjunction with Circuit 4-5, you can use a single power supply in common. Position Limit Switch 2 where the revolving solar array will physically contact it at the end of the solar day thereby closing its contacts and position. Position Limit Switch 1 where the revolving array will physically contact it at the dawn position thereby opening its contacts.

92 Circuits for Harvesting the Free Energy of the Sun

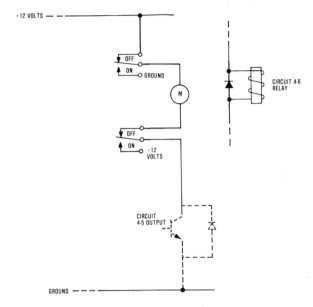

Figure 4-1. How to connect circuit 4-5 into Circuit 4-6.

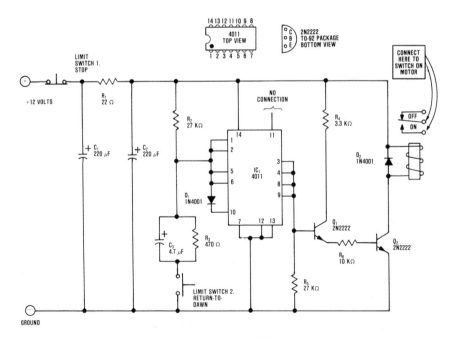

Circuit 4-6. Return-to-dawn control for your own solar array

Circuits for Harvesting the Free Energy of the Sun 93

PARTS LIST

C_1, C_2—220-μF, 25-volt aluminium electrolytic (Radio Shack 272-1017 or 272-1029)
C_3—4.7-μF, 25-volt aluminum electrolytic (Radio Shack 272-1012 or 272-1024)
D_1, D_2—1N4001 diodes (Radio Shack 276-1101)
IC_1—4011 integrated circuit (Radio Shack 276-2411)
Q_1, Q_2—2N2222 or similar transistors (Radio Shack 276-1617)
R_1—22 ohm, ¼-watt (Radio Shack 271-005)
R_2, R_5—27,000 ohm, ¼-watt (Radio Shack 271-1340)
R_3—470 ohm, ¼-watt (Radio Shack 271-1317)
R_4—3,300 ohm, ¼-watt (Radio Shack 271-1328)
R_6—10,000 ohm, ¼-watt (Radio Shack 271-1335)
Relay—12 volt, DPDT relay, 250 ohms or more coil resistance (Radio Shack 275-206)
Limit Switch$_1$—Normally closed pushbutton switch (Radio Shack 275-1548)
Limit Switch$_2$—Normally open pushbutton switch (Radio Shack 275-1547)

Circuit 4-7. Up/Down Control for Your Solar Array

This automatic control will increment your solar array up or down in order to adjust for changes in the sun's altitude as the day progresses and for seasonal variations. It is ideal for solar concentrators where energy must be accurately directed onto a boiler or a silicon solar panel.

In operation, if the solar array is pointed too high or too low, light entering a small enclosure box through a horizontal slot will strike either photoresistor R_1 or R_2 (respectively). Depending on which photoresistor is struck, the integrated circuit will activate motor M to run in a forward or reverse direction for a timed interval. This motor, in turn, increments the hinged solar array either up or down via a mechanical geared arrangement to correct the misalignment.

For construction, use 12-volt power supply 11-5. If you are using this circuit in conjunction with Circuits 4-5 and 4-6, they can all share a single supply. Position, "Light Level Adjust" potentiometer R_7, "# 1 ON Time Adjust" variable resistor R_{15}, and "# 2 ON Time Adjust" variable resistor R_{19} for easy access from the front of the circuit box. Install photoresistors R_1 and R_2 inside a small enclosure box which has a light-absorbent inner coating plus a horizontal slot to admit light. Position these photoresistors within the enclosure box for best effect, using a trial-and-error

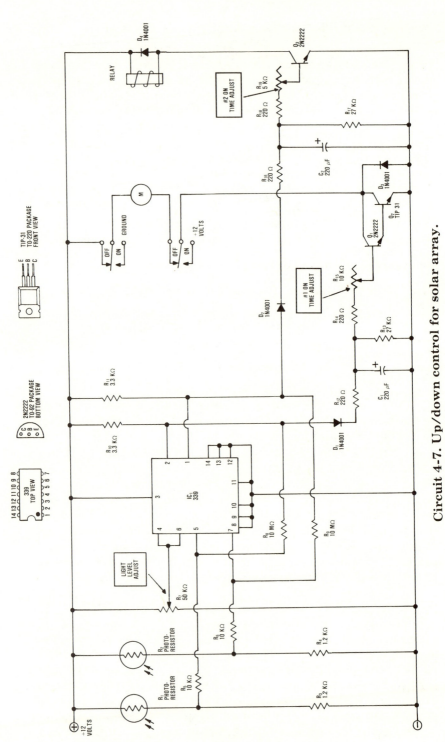

Circuit 4-7. Up/down control for solar array.

Circuits for Harvesting the Free Energy of the Sun 95

procedure. Next, connect the motor to the array with a geared-down mechanical arrangement so that as the motor spins, it slowly rotates the array either up or down (depending on which way the motor is spinning).

To adjust this circuit, shine a flashlight through the horizontal slot in the enclosure box onto photoresistor R_1, and then turn Light Level Adjust R_7 until the circuit just clicks the motor ON (if the motor rotates in the wrong direction, simply reverse the motor's connections). Next, adjust #1 ON Time Adjust R_{15} for sufficient motor ON time to increment the array a few degrees in the desired direction (but never so much that, in service, light will fall on photoresistor R_2). Finishing up, shine your light on photoresistor R_2, and adjust #2 ON Time Adjust R_{19} for sufficient motor ON time in the same manner.

PARTS LIST

C_1, C_2—220-μF 25-volt aluminum electrolytic (Radio Shack 272-1017 or 272-1029)
D_1, D_2, D_3, D_4—1N4001 diodes (Radio Shack 276-1101)
IC_1—339 integrated circuit (Radio Shack 276-1712)
M—Any 12-volt, low-current, reversible motor such as that from a scrap tape recorder
Q_1, Q_3—2N2222 or similar transistors (Radio Shack 276-1617)
Q_2—TIP 31 or similar transistor (Radio Shack 276-2017)
R_1, R_2—Cadmium sulfide photoresistors, about 500,000 ohms dark, 100 ohms light (Radio Shack 276-116)
R_3, R_4—1,200 ohm, ¼-watt (Radio Shack 271-024)
R_5, R_6—10,000 ohm, ¼-watt (Radio Shack 271-1335)
R_7—50,000 ohm, ¼-watt or more, linear taper potentiometer (Radio Shack 271-1716)
R_8, R_9—10,000,000 ohm, ¼-watt (Radio Shack 271-1365)
R_{10}, R_{11}—3,300 ohm, ¼-watt (Radio Shack 271-1328)
R_{12}, R_{14}, R_{16}, R_{18}—220 ohm, ¼-watt (Radio Shack 271-1313)
R_{13}, R_{17}—27,000 ohm, ¼-watt (Radio Shack 271-1340)
R_{15}—10,000 ohm, ¼-watt or more, linear taper potentiometer (Radio Shack 271-1715)
R_{19}—5,000 ohm, ¼-watt or more, linear taper potentiometer (Radio Shack 271-1714)
Relay—12-volt DPDT relay, 250 ohms or more coil resistance (Radio Shack 275-206)

In conclusion, although the prospects for the fossil fuels and other conventional sources of energy may appear bleak, solar energy is looking brighter than ever. Quite literally, the only costs involved are those of converting sunlight into more useful forms of energy. And inexpensive yet "intelligent" semiconductor circuits can play a vital role in reducing these costs to the point where energy is once more as abundant as it was in the 60's.

Ingenious Yet Simple Robot Circuits for Home, Shop, or Industry

There's just no good reason to perform many of those routine tasks around the home or workplace that tend to put our lives in such a rut. In many happy instances, a highly reliable semiconductor robot could faithfully handle those tasks at the cost of only a few cents worth of energy a day, thereby freeing our time for more worthwhile activities. This chapter will show you how to build a number of functional robot units which you can put to use right away.

For example, how often do you find yourself watching a clock for the purpose of turning OFF an appliance after some time interval is over? This is one chore that the Circuit 5-2 Electronic Switch With a Memory could easily handle. And because it has a "memory" which you program for an interval from about three seconds through six hours, it is useful for a wide range of applications.

Or is there some liquid storage tank or trough that you seem to be forever filling and re-filling? If so, why not assign this task to the Circuit 5-3 Tank Filling Robot. It will work day and night and even on the weekends to maintain the liquid level in the tank to within whatever limits you specify.

Because many routine jobs must be performed at dusk (or dawn), we've included a circuit which activates with decreasing (or increasing) light. The Circuit 5-4 Relay Controller Which "Sees" closes a set of electrical contacts at dusk for a timed interval thereby completing a job which you would otherwise have to tend—from switching ON and then OFF lights, to watering your lawn. And with an adjustable timing range from seconds through hours, it can perform a wide assortment of tasks.

There are even several comparatively "smart" robots which you can put to work on fairly complex jobs. They will sequentially close relays for adjustable time intervals to complete a task in a step-by-step fashion (and when you think about it, most jobs follow a step-by-step procedure). The Circuit 5-6 Tireless 1-2-1-2 Robot can successfully complete a two-step function... and then tirelessly do this function again and again until you "tell" it to stop. And the Circuit 5-7 1-2-3 Robot can handle almost any task which you can break down into discrete steps performed in sequential order.

Circuit 5-1. Precise-Temperature Water Heater

This circuit not only brings liquid up to whatever temperature you select, but trickles through just the right amount of energy to hold it at that temperature indefinitely. It is ideal for a wide range of industrial and home uses such as chemical and photographic processing, aquarium heating—even simmering soup. As an important added bonus, this Triac-based circuit regulates power flow with virtually no waste.

For construction, first remember that you will be dealing with high voltage throughout most of this circuit and act accordingly. Mount components inside a circuit box with front-panel access for "Temperature Adjust" variable resistor R_3, an input connection for remotely located thermistor R_4, and an output socket for the electrical heating element (such as a hot plate). For easy temperature adjustment, install a pointer-type knob on the shaft of Temperature Adjust R_3. Be sure to select a Triac with sufficient current capability for your intended use.

To calibrate and operate this circuit, first be sure to fully immerse heat-sensing thermistor R_4 in the liquid to be heated, avoiding contact with the sides of the container in order to avoid spurious results. Next, make several trial settings of Temperature Adjust R_3 while monitoring liquid temperature with an accurate thermometer. At each setting, allow the liquid to reach and hold some temperature and then mark this value on the circuit faceplate opposite the R_3 pointer. Repeat until you have covered the range of temperatures that you wish to use.

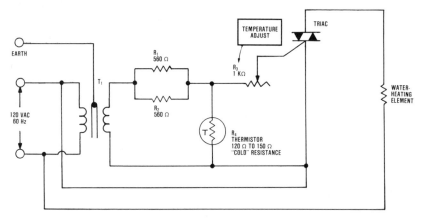

Circuit 5-1. Precise-temperature water heater

Simple Robot Circuits for Home, Shop, or Industry 101

PARTS LIST

R_1, R_2—560 ohm, ½-watt (Radio Shack 271-020)
R_3—1,000 ohm, ¼-watt (or more), linear taper potentiometer
R_4—120-ohm to 150-ohm "cold" resistance thermistor
T_1—12-volt, 300-mA to 500-mA secondary, miniature transformer (Radio Shack 273-1385)
Triac—Any Triac with sufficient current capability to drive heater element (such as Radio Shack 276-1001)

Circuit 5-2. Electric Switch With a Memory

Why be a clockwatcher when this circuit will "remember" to turn OFF an appliance after any time interval you select—from seconds through hours? It can activate a darkroom timer for a precisely adjustable interval . . . turn OFF the heat at just the right time for cooking . . . or regulate a wide variety of industrial processes such as mixing, heating, and sanding. Truly one of the most versatile and useful semiconductor circuits anywhere!

For construction, use 12-volt supply 11-4 (adjustable) or 11-12. If possible, use the adjustable supply and set it for 13 volts in order to provide extra voltage for closing the relay. Choose the proper value for timing capacitor C_1 to suit your purpose from the included chart. Mount "Standby/Timed Interval" switch S_1, "Fine ON Time Adjust" variable resistor R_4, and "Coarse ON Time Adjust" variable resistor R_5 for access from the front of the circuit box. For easy ON time adjustment, install pointer-type knobs on the shafts of R_4 and R_5. Choose a relay with contacts of sufficient current capability for your intended application.

To operate this circuit, begin with Standby/Timed Interval switch S_1 in the Standby position. This is to charge timing capacitor C_1. Then when you click this switch to the Timed Interval position, the relay will close for an interval switching ON an appliance. Make various settings of ON time adjust R_4 and R_5, marking the setting on the front of the circuit box opposite the pointer knobs for later reference.

PARTS LIST

C_1—25-volt, aluminum electrolytic (see chart for approximate timing range). (Check Radio Shack shelves)
D_1—IN4001 diode (Radio Shack 276-1101)
Q_1, Q_2, Q_3—2N2222 or similar transistors (Radio Shack 276-1617)
R_1—47 ohm, ¼-watt (Radio Shack 271-1307)
R_2—3,300 ohm, ¼-watt (Radio Shack 271-1328)

102 *Simple Robot Circuits for Home, Shop, or Industry*

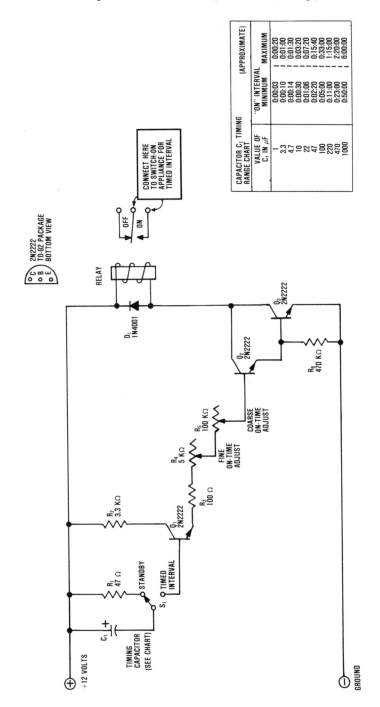

Circuit 5-2. Electric switch with a memory

Simple Robot Circuits for Home, Shop, or Industry 103

R_3—100 ohm, ¼-watt (Radio Shack 271-1311)
R_4—5,000 ohm, ¼-watt (or more), linear taper potentiometer (Radio Shack 271-1714)
R_5—100,000 ohm, ¼-watt (or more), linear taper potentiometer (Radio Shack 271-092)
R_6—470,000 ohm, ¼-watt (Radio Shack 271-1354)
Relay—SPDT, 12-volt relay, 160 ohms or more coil resistance (such as Radio Shack 275-218)
S_1—SPDT switch (Radio Shack 275-613)

Circuit 5-3. Tank-Filling Robot

No need to waste your valuable time on the routine job of filling and refilling a liquid storage tank when this robot circuit can take over your responsibility for only a few pennies' worth of energy a day. Simply position Sensor Probe₁ at the minimum liquid level you wish to maintain ... Sensor Probe₂ at the maximum ... and this circuit will tirelessly switch ON and OFF an electric pump or solenoid valve to maintain these limits. Innumerable applications with fairly conductive liquid includes use for a well-to-attic tank pump-control, for filling livestock watering troughs—even as a highly reliable toilet tank filler.

For construction, use 12-volt supply 11-4 (adjustable) or 11-12. If possible, use the adjustable supply and set it for 13 volts in order to provide extra voltage for closing the relay. Set perfboard—mounted "Sensitivity Adjust" variable resistors R_3 and R_4 to their mid-positions (only in rare installations is it necessary to set them for more or less sensitivity). Locate the Negative Probe at a low point in the tank with Sensor Probe₁ and Sensor Probe₂ freely suspended directly above. Almost any metallic objects, such as bolts, can serve as probes. Be sure to locate the circuit box in a dry, protected location to keep it out of harm's way.

PARTS LIST

D_1, D_2—IN4001 diodes (Radio Shack 276-1101)
IC_1—4011 integrated circuit (Radio Shack 276-2411)
Q_1, Q_2—2N2222 or similar transistors (Radio Shack 276-1617)
R_1, R_2, R_6—27,000 ohm, ¼-watt (Radio Shack 271-1340)
R_3, R_4—1,000,000 ohm, ¼-watt (or more) linear taper potentiometer (Radio Shack 271-229)
R_5—1,000 ohm, ¼-watt (Radio Shack 271-1321)
R_7—3,300 ohm, ¼-watt (Radio Shack 271-1328)
R_8—10,000 ohm, ¼-watt (Radio Shack 271-1335)
Relay—SPDT, 12-volt relay, 160 ohms or more coil resistance (Radio Shack 275-218)

104   Simple Robot Circuits for Home, Shop, or Industry

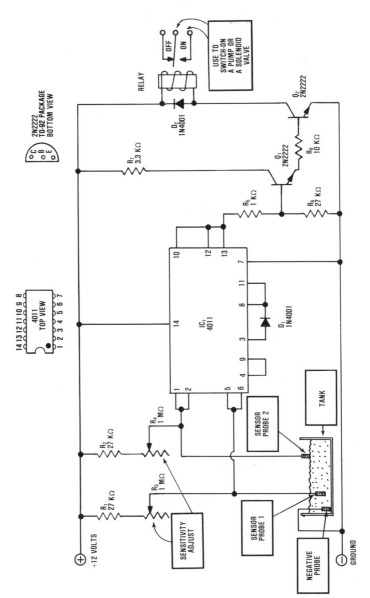

Circuit 5-3. Tank-filling robot

Circuit 5-4. Relay Controller That "Sees"

This circuit acts like a hired hand around your home or workplace to take care of chores that must be performed at dusk (or dawn). As ambient light decreases in the evening, it activates a relay to switch ON an appliance for an adjustable timed interval—from seconds through hours. You can have it water your lawn or garden via a solenoid valve, turn ON a sign at your place of work, even turn ON your front porch light to welcome your family home. To have this circuit activate at dawn rather than at dusk, simply reverse the position of R_1 and photoresistor R_2.

For construction, use 12-volt supply 11-4 (adjustable) or 11-12. If possible, use the adjustable supply and set it for 13 volts in order to provide extra voltage for closing the relay. Select a C_1 timing capacitor from the included chart to approximate your required ON period. Mount "Trigger Level Adjust" potentiometer R_3 and "ON Time Adjust" variable resistor R_{11} for front panel access. For easy ON time adjustment, install a pointer knob on the shaft of R_{11}. Be sure to position photoresistor R_2 so that it accurately reads ambient light conditions and cannot be falsely triggered by shadows, automobile headlamps, etc. Select a relay with a contact rating adequate for your intended application.

For the initial adjustment, energize the circuit and rotate Trigger Level Adjust R_3 until the relay is activated. Next, use a trial-and-error procedure to set ON Time Adjust R_{11} for the desired time period, making calibration marks on the circuit box opposite the pointer knob for easy later reference. Finishing up, wait until dusk (or dawn) and then rotate R_3 until the relay just clicks ON. Henceforth, the circuit will close the relay daily at this light level.

PARTS LIST

C_1—25-volt, aluminum electrolytic (see chart for approximate timing range) (check Radio Shack shelves)
D_1, D_2—IN4001 diodes (Radio Shack 276-1101)
IC_1—339 integrated circuit (Radio Shack 276-1712)
Q_1, Q_2, Q_3, Q_4,—2N2222 or similar transistors (Radio Shack 276-1617)
R_1—1,200 ohm, ¼-watt (Radio Shack 271-024)
R_2—Cadmium sulfide photoresistor (about 500,000 ohms dark, 100 ohms light) (Radio Shack 276-116)

106 Simple Robot Circuits for Home, Shop, or Industry

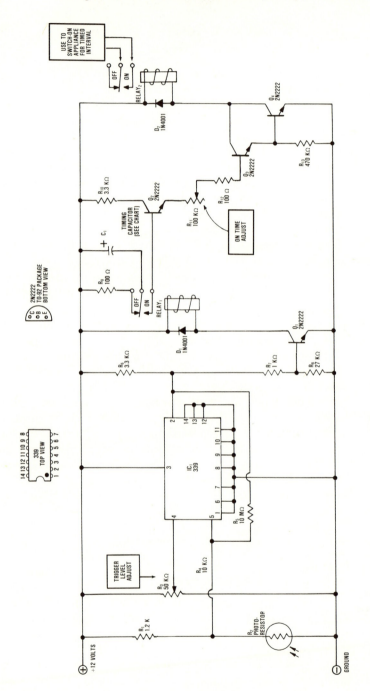

Circuit 5-4. Relay controller that "sees"

Simple Robot Circuits for Home, Shop, or Industry   107

R_3—50,000 ohm, ¼-watt (or more) linear taper potentiometer (Radio Shack 271-1716)
R_4—10,000 ohm, ¼-watt (Radio Shack 271-1335)
R_5—10,000,000 ohm, ¼-watt (Radio Shack 271-1365)
R_6, R_{10}—3,300 ohm, ¼-watt (Radio Shack 271-1328)
R_7—1,000 ohm, ¼-watt (Radio Shack 271-1321)
R_8—27,000 ohm, ¼-watt (Radio Shack 271-1340)
R_9, R_{12}—100 ohms, ¼-watt (Radio Shack 271-1311)
R_{11}—100,000 ohm, ¼-watt (or more), linear taper potentiometer (Radio Shack 271-092)
R_{13}—470,000 ohm, ¼-watt (Radio Shack 271-1354)
$Relay_1$, $Relay_2$—SPDT, 12-volt relay, 160 ohms or more coil resistance (such as Radio Shack 275-206 or 275-218)

CAPACITOR C_1 TIMING RANGE CHART (APPROXIMATE)		
VALUE OF C_1 IN µF	"ON" INTERVAL MINIMUM	MAXIMUM
1	0:00:03	0:00:20
3.3	0:00:10	0:01:00
4.7	0:00:14	0:01:30
10	0:00:30	0:03:20
22	0:01:06	0:07:20
47	2:02:20	0:15:40
100	0:05:00	0:33:00
220	0:11:00	1:15:00
470	0:23:00	2:20:00
1000	0:50:00	6:00:00

Circuit 5-5. Relay Controller That "Hears"

This circuit constantly "listens," and when sound volume exceeds a pre-set level, it closes a relay for an adjustable time interval thereby switching ON or OFF an appliance. It can stand watch over machinery, and signal you if any unusual loud noise should occur ... listens in at the nursery ... opens a pet entrance in response to a bark ... even opens your garage door in response to a honk.

As used for voice command, this circuit can operate tools, thereby freeing one of your hands for other purposes. In some cases, it can allow one person to complete a job where two persons were needed before. At full sensitivity it will activate with a normal tone of voice at 3 feet, a shout at 10 feet, and much farther when the microphone is used in conjunction with a parabolic reflector.

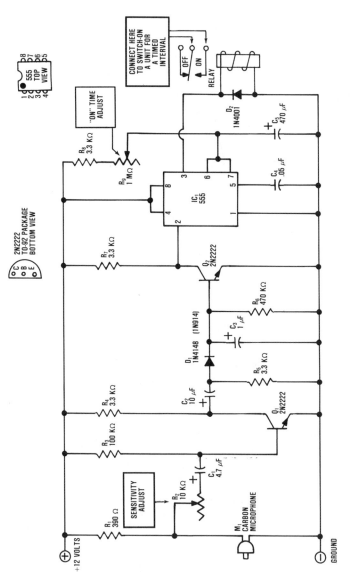

Circuit 5-5. Relay controller that "hears"

Simple Robot Circuits for Home, Shop, or Industry 109

For construction, use 12-volt supply 11-4 (adjustable) or 11-12. If possible, use the adjustable supply and set it for 13 volts in order to provide extra voltage for closing the relay. Mount "Sensitivity Adjust" variable resistor R_2 and "ON Time Adjust" variable resistor R_9 for front-panel access. Equip R_9 with a pointer-type knob for easy ON time adjustment. Try to plug this circuit into an electrical outlet which is not on the same fuse as transient-producing gear such as fluorescent lighting or electric motors so as to avoid false triggering. Ensure that the relay you select has contacts adequate for the appliance to which they are to be connected.

PARTS LIST

C_1—4.7-μF, 25-volt aluminum electrolytic (Radio Shack 272-1012 or 272-1024)
C_2—10-μF, 25-volt aluminum electrolytic (Radio Shack 272-1013 or 272-1025)
C_3—1-μF, 16-volt aluminum electrolytic (Radio Shack 272-1419)
C_4—0.05-μF, 50-volt ceramic disk (Radio Shack 272-134)
C_5—470-μF, 25-volt aluminum electrolytic (Radio Shack 272-1018 or 272-1030)
D_1—1N4148 (1N914) (Radio Shack 276-1122)
D_2—1N4001 (Radio Shack 276-1101)
IC_1—555 integrated circuit (Radio Shack 276-1723)
M_1—Carbon microphone (the mouthpiece from an old telephone works well).
Q_1, Q_2—2N2222 or similar transistors (Radio Shack 276-1617)
R_1—390 ohm, ¼-watt (Radio Shack 271-018)
R_2—10,000 ohm, ¼-watt (or more) linear taper potentiometer (Radio Shack 271-1715)
R_3—100,000 ohm, ¼-watt (Radio Shack 271-1347)
R_4, R_5, R_7, R_8—3,300 ohm, ¼-watt (Radio Shack 271-1328)
R_6—470,000 ohm, ¼-watt (Radio Shack 271-1354)
R_9—1,000,000 ohm, ¼-watt (or more), linear taper potentiometer (Radio Shack 271-211)
Relay—SPDT, 12-volt relay, 160 ohms or more coil resistance (such as Radio Shack 275-206 or 275-218)

Circuit 5-6. Tireless 1-2-1-2 Robot

An electronic "brain" which will sequentially switch ON alternate appliances for timed intervals to complete a task... and repeatedly do this task until you "tell" it to stop. For example, $Relay_1$ can turn ON a conveyor belt dropping a measured quantity of vegetables into a shredder. Then when $Relay_1$ drops out, $Relay_2$

110   Simple Robot Circuits for Home, Shop, or Industry

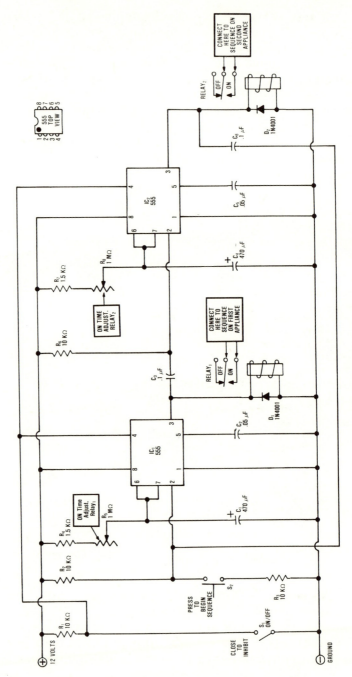

Circuit 5-6. Tireless 1-2-1-2 robot

can switch on the shredder to complete the task. This sequence will then repeat indefinitely. ON time for each relay is continuously adjustable from a few seconds through about 6 minutes.

To construct this circuit, use 12-volt supply number 11-4 (adjustable) or 11-12. If possible, use the adjustable supply and set it for 13 volts in order to provide extra voltage for closing the relay. Position "Close TO Inhibit" switch S_1, "Press to Begin Sequence" switch S_2, and "ON Time Adjust" variable resistors R_5 and R_8 for easy access from the front of the circuit box. Be sure to include pointer-type knobs for R_5 and R_8 to facilitate adjustment. Select relays with sufficient current-switching capabilities for the task.

For installation and operation, try to plug this circuit into an electrical outlet on a different fuse from such transient-producing gear as fluorescent lights and electric motors. Also, remember that high voltage will be present at the relays, and act accordingly. To operate this circuit, first shut Close To Inhibit switch S_1, and then energize the circuit. Next, open S_1 and momentarily push Press To Begin Sequence switch S_2, in order to start the 1-2-1-2 function. Adjust ON time for Relay$_1$ and Relay$_2$ with R_5 and R_8 respectively.

To reset or temporarily deactivate this circuit, simply close switch S_1.

PARTS LIST

C_1, C_4—470-μF, 25-volt aluminum electrolytic (Radio Shack 272-1018 or 272-1030)
C_2, C_5—0.05-μF, 50-volt ceramic disk (Radio Shack 272-134)
C_3, C_6—0.1-μF, 50-volt ceramic disk (Radio Shack 272-135)
D_1, D_2—1N4001 diodes (Radio Shack 276-1101)
IC_1, IC_2—555 integrated circuits (Radio Shack 276-1723)
R_1, R_2, R_3, R_6,—10,000 ohm, ¼-watt (Radio Shack 271-1335)
R_4, R_7,—1,500 ohm, ¼-watt (Radio Shack 271-025)
R_5, R_8,—1,000,000 ohm, ¼-watt (or more), linear taper potentiometers (Radio Shack 271-211)
Relay$_1$, Relay$_2$—SPDT, 12-volt relays, 160 ohms or more coil resistance (such as Radio Shack 275-206 or 275-218)
S_1—ON/OFF switch (Radio Shack 275-602)
S_2—Normally open pushbutton switch (Radio Shack 275-1547)

Circuit 5-7. 1-2-3 ... Robot

This electronic "brain" can complete any task that can be broken down into discrete steps performable by closing relays for timed periods in sequential order. And you can add as many relay

112 Simple Robot Circuits for Home, Shop, or Industry

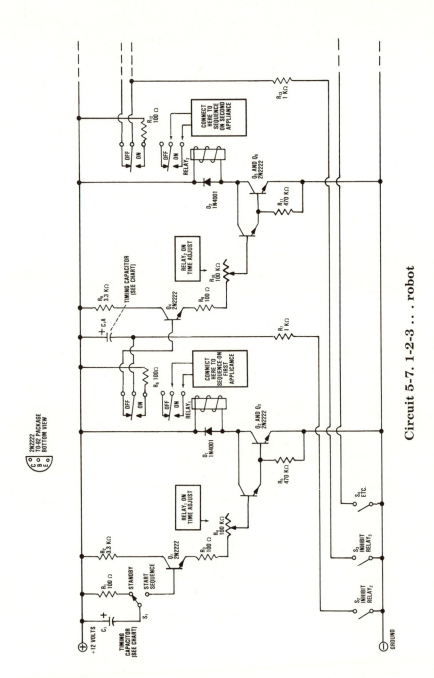

Circuit 5-7. 1-2-3 .. robot

Simple Robot Circuits for Home, Shop, or Industry

CAPACITOR C_1 TIMING RANGE CHART (APPROXIMATIONS)		
VALUE OF C_1 IN μF	"ON" INTERVAL MINIMUM	MAXIMUM
1	0:00:03	0:00:20
3.3	0:00:10	0:01:00
4.7	0:00:14	0:01:30
10	0:00:30	0:03:20
22	0:01:06	0:07:20
47	0:02:20	0:15:40
100	0:05:00	0:33:00
220	0:11:00	1:15:00
470	0:23:00	2:20:00
1000	0:50:00	6:00:00

stages as you need in order to complete even complex tasks. For example, to use this circuit to automatically mix cement, Relay$_1$ can activate a conveyor belt thereby depositing a measured amount of cement into a mixing drum. Relay$_2$ can next open a solenoid valve admitting the right quantity of water. Relay$_3$ can then turn ON the mixing motor, and Relay$_4$ empty the drum and signal the operator. The operator then inspects the completed job, and re-cycles "Standby/Start Sequence" switch S_1 to begin the next batch.

For construction, use 12-volt supply number 11-4 (adjustable) or 11-12. If possible, use the adjustable supply and set it for 13 volts in order to provide extra voltage for closing the relays. Select a C_1 timing capacitor for each relay stage for the required ON time from the included chart. Position "Standby/Start Sequence" switch S_1, "Relay ON Time Adjust" variable resistors R_4, R_{10}, etc., and "Inhibit Relay$_{2\ 3\ 4\ etc.}$" switches for front-panel access. To add additional relay stages, simply continue to build in the parts sequence which follows Relay$_2$. Select relays with sufficient current-handling abilities for the task.

To operate this circuit, first close Inhibit switches $S_{2\ 3\ 4\ etc.}$, and then click Standby/Start Sequence switch S_1 to the Standby position. Next, energize the circuit and open $S_{2\ 3\ 4\ etc.}$. To begin sequential operation, simply click Standby/Start Sequence switch S_1 to the Start Sequence position. Set ON time for each relay stage with the respective adjustment potentiometer using a trial-and-error procedure.

To re-cycle this circuit once the operation is complete, simply click S_1 to the Standby position for one second or more, and then return it to the Start Sequence position.

PARTS LIST

C_1, C_{1A}, (etc.)—25-volt, aluminum electrolytic (see chart for approximate timing range) (check Radio Shack shelves)
D_1, D_2—1N4001 diodes (Radio Shack 276-1101)
Q_1, Q_2, Q_3, Q_4, Q_5, Q_6,—2N2222 or similar transistors (Radio Shack 276-1617)
R_1, R_3, R_6, R_9, R_{12},—100 ohm, ¼-watt (Radio Shack 271-1311)
R_2, R_8—3,300 ohm, ¼-watt (Radio Shack 271-1328)
R_4, R_{10}—100,000 ohm, ¼-watt (or more), linear taper potentiometers (Radio Shack 271-092)
R_5, R_{11}—470,000 ohm, ¼-watt (Radio Shack 271-1354)
R_7, R_{13}—1,000 ohms, ¼-watt (Radio Shack 271-1321)
$Relay_1$, $Relay_2$—DPDT, 12-volt relays, 160 ohms or more coil resistance (such as Radio Shack 275-206 or 275-218)
S_1—SPDT switch (Radio Shack 275-613)
S_2, S_3, S_4 (etc.)—ON/OFF switches (Radio Shack 275-602)

Wrapping things up, the realities of the 80's will dictate that we must work smarter—not just harder. And in the same way that semiconductor circuits can now handle routine mathematical calculations which we previously had to laboriously perform by hand, they can take over boring jobs that presently take up so much of our valuable time. What is more, you can get a head start on the future with the semiconductor circuits described in this chapter.

How to Use Inexpensive Semiconductors for Top-Quality Sound Reproduction

The development of integrated circuit audio chips has brought real benefits to the technician. Because almost all the components which make up an audio amplifier can be housed within the chip, circuit assembly is quick and interconnection problems relatively rare. What's more, the cost of the IC's which do such wonderful things is low—often less than the price of a single discrete component. To find out how you can explore the latest in IC sound chips, just read on ...

One welcome addition to almost any living area or work area is "pink noise" generator Circuit 6-1. It produces soothing background noise to help you relax while you work or play. Pink noise, of course, is random sound whose amplitude progressively decreases with frequency. Natural generators of such noise which this circuit can imitate include rain-on-the-roof, wind-through-the-trees, and ocean surf.

And for saving time, there's hardly a circuit which can compete with the Handy 2-Way Intercom of Circuit 6-2. It can be worth its weight in gold almost anywhere that you have to communicate between two (or more) places, such as from front desk to stockroom. To calculate the total time that this circuit can save you, just add up the minutes you spend every day walking between these two points.

Or perhaps you could benefit from a sound-boost unit for an existing audio set ... or a PA/loudhailer system. If so, Circuit 6-3 or Circuit 6-4 could be the easy-to-assemble answer to your needs.

The instructions in this chapter will also show you how to snap together a complete stereo amplifier with surprisingly few parts. The stereo phonograph/tape recorder amplifier depicted in Circuit 6-5 can be ideal for your room, apartment, or office. And with an output of two watts per channel, it provides plenty of volume for normal listening purposes.

For the true audiophile, there are circuits which can pick up the performance of your existing equipment. One good example of this is the Super Stereo Preamp of Circuit 6-6, which will improve signal-to-noise ratio and reduce distortion in your main amplifier. Or the Hi-Fi Magnetic Cartridge Preamp of Circuit 6-7, which lets you add the superior performance of a magnetic phonograph cartridge to a stereo amplifier having only a ceramic phono or auxiliary input. And for mixing your own music, Circuit 6-8 gives you the advantage of a professional unit at a price you can easily afford. Four independent inputs plus a stereo/mono switch lets you precisely blend sound to best advantage.

Circuit 6-1. Restful Waterfall and Ocean Background Sounds On A Chip

Simply connect this circuit to a speaker for soothing, commercial-free "pink noise" background sounds. A real plus for home, shop, or office. Rotary switch S_1 allows you to vary the tone to suit your mood—from a sound almost indistinguishable from the splatter of rain on a tin roof, to heavy ocean surf. Very pleasing effects.

For building this audio circuit, use 15-volt power supply number 11-6 or 11-13. Mount 6-position rotary switch, and "Volume Adjust" potentiometer R_2 for easy access from the front of the circuit box. Remember that—as with almost all audio circuits—a good, quality speaker is important for best sound rendition.

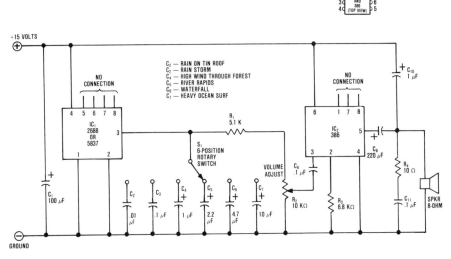

Circuit 6-1. Restful waterfall and ocean background sounds on a chip

PARTS LIST

C_1—100-μF, 25-volt aluminum electrolytic (Radio Shack 272-1016 or 272-1028)
C_2—0.01-μF, 50-volt ceramic disk (Radio Shack 272-131)
C_3, C_8, C_{11}—0.1-μF, 50-volt ceramic disk (Radio Shack 272-135)
C_4, C_{10}—1-μF, 25-volt solid tantalum (Radio Shack 272-1419

C_5—2.2-μF, 25-volt solid tantalum (Radio Shack 272-1420)
C_6—4.7-μF, 25-volt solid tantalum (Radio Shack 272-1422)
C_7—10-μF, 25-volt solid tantalum (Radio Shack 272-1423)
C_9—220-μF, 25-volt aluminum electrolytic (Radio Shack 272-1017 or 272-1029)
IC_1—2688 or 5837 integrated circuit (Radio Shack 276-1768)
IC_2—386 integrated circuit (Radio Shack 276-1731)
R_1—5,100 ohm, ¼-watt (Radio Shack 271-1330 is OK)
R_2—10,000 ohm, ¼-watt (or more) audio taper potentiometer (Radio Shack 271-1721)
R_3—6,800 ohm, ¼-watt (Radio Shack 271-1333)
R_4—10 ohm, ¼-watt (Radio Shack 271-1301)
S_1—6-position make-before-break rotary switch (Radio Shack 275-1385 is OK)
SPKR—Any 8-ohm speaker

Circuit 6-2. Handy 2-way Intercom

A complete intercom system which will save you many steps every day! You can use it at home or at work to communicate between kitchen and workshop... upstairs to downstairs... front desk to stockroom... or to answer the doorbell. You can even use it

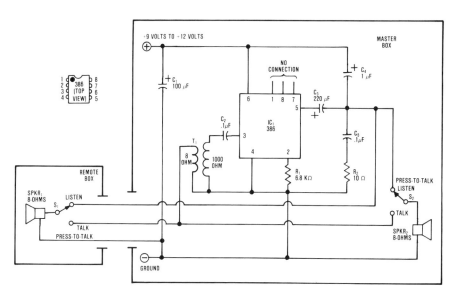

Circuit 6-2. Handy 2-way intercom

to listen in at the nursery by taping down the press-to-talk button. Features almost static-free operation with a tone so crisp that you can use it to pipe in music. Consumes so little power that you can leave it on continuously.

For construction tips, use 9-volt or 12-volt power supply number 11-4, 11-11, or 11-12. Note that almost all the circuitry is housed in the master box, which is located in a protected area near an electrical outlet. The remote box contains only a speaker and a switch: OK to shunt in additional remote boxes. To interconnect the boxes, use 3-conductor intercom-type wire, keeping away from power conductors as much as possible in order to minimize hum. Be sure to connect the SPDT momentary-contact press-to-talk switches S_1 and S_2 so that they are normally in the "Listen" position.

PARTS LIST

C_1—100-μF, 25-volt aluminum electrolytic (Radio Shack 272-1016 or 272-1028)
C_2, C_5—0.1-μF, 50-volt ceramic disk (Radio Shack 272-135)
C_3—220-μF, 25-volt aluminum electrolytic (Radio Shack 272-1017 or 272-1029)
C_4—1-μF, 25-volt solid tantalum (Radio Shack 272-1419)
IC_1—386 integrated circuit (Radio Shack 276-1731)
R_1—6,800 ohm, ¼-watt (Radio Shack 271-1333)
R_2—10 ohm, ½-watt (Radio Shack 271-001)
S_1, S_2—SPDT momentary-contact pushbutton switches (Radio Shack 275-1549)
$SPKR_1$, $SPKR_2$—Any 8-ohm magnetic-type speakers.
T_1—8-ohm: 1000-ohm miniature audio output transformer (Radio Shack 275-1380)

Circuit 6-3. Extra Sound for Any Low-Powered Audio System

Here's a quick and effective way to boost sound from any low-powered radio, television, movie projector, intercom, or other audio set. Drives up to three 8-ohm speakers connected in parallel with low distortion. No need to worry about damage to this circuit due to overloading, as it automatically shuts itself down in the event that thermal dissipation becomes too great. It features complete output short circuit protection also.

For assembly, use supply 11-5 set up for 12-volt operation. Be certain to equip the 383 IC with a good heat sink because distortion temporarily occurs if the internal thermal overload mechanism is activated. When used with multiple speakers, be

Inexpensive Semiconductors for Top-Quality Sound 121

sure that all speakers are correctly wired to push and pull in unison. Otherwise loss of sound through "phase cancellation" will result. Mount "Volume Adjust" potentiometer R_3 for front panel access.

To tap off a signal from the audio set for amplification, simply jump its speaker terminals with alligator clips (does not interfere with operation). From this boost circuit, use standard speaker wire to route the amplified signal to the additional speakers. Set Volume Adjust R_3 for the desired amount of loudness.

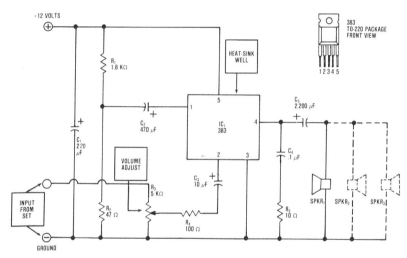

Circuit 6-3. Extra sound for any low-powered audio set

PARTS LIST

C_1—220-μF, 25-volt aluminum electrolytic (Radio Shack 272-1017 or 272-1029)
C_2—470-μF, 25-volt aluminum electrolytic (Radio Shack 272-1018 or 272-1030)
C_3—10-μF, 25-volt aluminum electrolytic (Radio Shack 272-1013 or 272-1025)
C_4—0.1-μF, 50-volt ceramic disk (Radio Shack 272-135)
C_5—2,200-μF, 25-volt aluminum electrolytic (Radio Shack 272-1020)
IC_1—383 integrated circuit (Radio Shack 276-703)
R_1—1,800 ohm, ¼-watt (Radio Shack 271-1324)
R_2—47 ohm, ¼-watt (Radio Shack 271-1307)
R_3—5,000 ohm, ¼-watt (or more), audio taper potentiometer (Radio Shack 271-1720)
R_4—100 ohm, ¼-watt (Radio Shack 271-1311)
R_5—10 ohm, ½-watt (Radio Shack 271-001)
$SPKR_1$, $SPKR_2$, $SPKR_3$—Any 8-ohm speakers

Circuit 6-4. PA/Megaphone

Ideal for sporting events, festive occasions, or for paging. It can be operated from a line-powered supply, or from any 12-volt automotive-type battery. Drives up to three standard or horn-type speakers with up to a total of 8-watts of output power. Gives a good, clean tone with virtually none of the buzz usually associated with low-cost systems. Includes automatic thermal shutdown and complete output short circuit protection.

For assembly and operation, you can energize this circuit with 12-volt power supply number 11-5, or else connect it directly to the positive and negative poles of a 12-volt automotive-type battery using jumper clips. No need to remove the battery from the vehicle. Connecting into any other portion of the vehicle's wiring system other than directly to the positive and negative battery poles is not recommended because of the possibility of damage due to transient voltages. Be sure to heat-sink the 383 IC well to avoid temporary distortion and loss of volume caused by thermal overload shutdown. For the required microphone, you can use either an 8-ohm, magnetic-type speaker coupled to an 8 ohm: 1000 ohm miniature audio transformer (good) ... or else a low-impedance dynamic microphone (better). Take care to separate the microphone from the output speaker(s) in order to avoid regenerative whine. Position "Volume Adjust" potentiometer R_2 for front panel access.

PARTS LIST

C_1, C_3, C_9—0.1-μF, 50-volt ceramic disk (Radio Shack 272-135)
C_2—100-μF, 25-volt aluminum electrolytic (Radio Shack 272-1016 or 272-1028)
C_4, C_5—470-μF, 25-volt aluminum electrolytic (Radio Shack 272-1018 or 272-1030)
C_6—10-μF, 25-volt aluminum electrolytic (Radio Shack 272-1013 or 272-1025)
C_7—2,200-μF, 16-volt aluminum electrolytic (Radio Shack 272-1020)
C_8—1-μF, 25-volt solid tantalum (Radio Shack 272-1419)
IC_1—386 integrated circuit (Radio Shack 276-1731)
IC_2—383 integrated circuit (Radio Shack 276-703)
M_1—Low-impedance dynamic microphone
R_1—6,800 ohm, ¼-watt (Radio Shack 271-1333)
R_2—5,000 ohm, ¼-watt (or more) audio taper potentiometer (Radio Shack 271-1720)
R_3—100 ohm, ¼-watt (Radio Shack 271-1311)
R_4—1,800 ohm, ¼-watt (Radio Shack 271-1324)

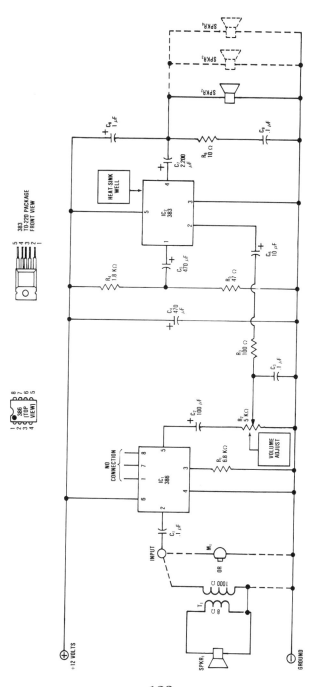

Circuit 6-4. PA/Megaphone/Loudhailer system

123

R_5—47 ohm, ¼-watt (Radio Shack 271-1307)
R_6—10 ohm, ½-watt (Radio Shack 271-001)
$SPKR_1$—Any 8-ohm, magnetic-type speaker
$SPKR_2$, $SPKR_3$, $SPKR_4$—Any 8-ohm speakers
T_1—8-ohm: 1,000-ohm miniature audio output transformer (Radio Shack 273-1380)

Circuit 6-5. A High-Fidelity Stereo Phonograph/Tape Recorder Amplifier Need Not Be Expensive

The plans for this high-performance yet low-cost circuit can alone be worth the price of this book. It's great for use in your room, apartment, office, or shop. Two watts of output power per channel provide plenty of power for "bookshelf"-sized speakers or stereo headphones. Tone control attenuates bass at one extreme ... treble on the other. Performance specifications include a total harmonic distortion figure of about .05% at low volume, a ripple rejection ratio of 70 dB, and a channel separation figure of 75 dB. Built-in current limiting and thermal overload shutdown eliminate damage from most normal types of abuse.

For construction, use 15-volt supply 11-6 or 11-13. If possible, use the 1877 IC as it is somewhat superior to the 377. Some form of heat sink is advisable. To minimize hum, try to position circuit components away from the power supply components. Also, use shielded-type audio cable (with the outer braid attached to circuit ground) for routing the input signals between the input jacks and IC pins. Earthing the transformer case may help too. Position "Volume Adjust" potentiometers R_2 and R_3 plus "Tone Adjust" potentiometers R_4 and R_5 for front-panel access.

For additional tips, use good-quality speakers for best sound performance. If you install a stereo headphone jack, remember to solder a 22 ohm to 33 ohm, ½-watt resistor in series with each headphone speaker to prevent overload.

PARTS LIST

C_1—1,000-μF, 35-volt aluminum electrolytic (Radio Shack 272-1019 or 272-1032)
C_2, C_3, C_4, C_5—0.1-μF, 50-volt ceramic disk (Radio Shack 272-135)
C_6—220-μF, 25-volt aluminum electrolytic (Radio Shack 272-1017 or 272-1029)
C_7, C_8—470-μF, 25-volt aluminum electrolytic (Radio Shack 272-1018 or 272-1030) (con't)

Inexpensive Semiconductors for Top-Quality Sound 125

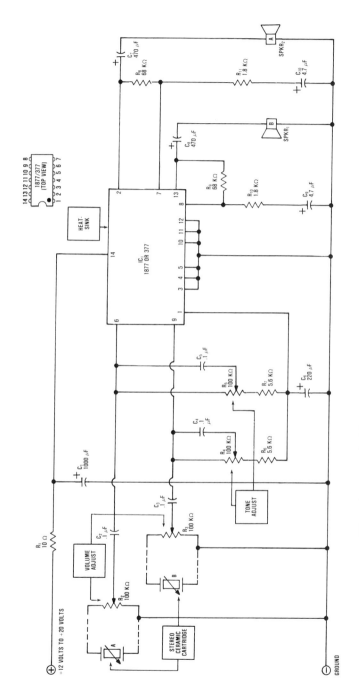

Circuit 6-5. A high-fidelity stereo phonograph/tape recorder amplifier need not be expensive

C_9, C_{10}—4.7-μF, 25-volt aluminum electrolytic (Radio Shack 272-1012 or 272-1024)
IC_1—1877 or 377 integrated circuit (Radio Shack 276-702)
R_1—10 ohm, ½-watt (Radio Shack 271-001)
R_2, R_3—100,000 ohm, ¼-watt (or more), audio taper potentiometer (Radio Shack 271-1722)
R_4, R_5—100,000 ohm, ¼-watt (or more), linear taper potentiometer (Radio Shack 271-092)
R_6, R_7—5,600 ohm, ¼-watt (Radio Shack 271-031)
R_8, R_9—68,000 ohm, ¼-watt (Radio Shack 271-1345)
R_{10}, R_{11}—1,800 ohm, ¼-watt (Radio Shack 271-1324)
$SPKR_1$, $SPKR_2$—Any good quality 8-ohm speakers

Circuit 6-6. Super Stereo Preamp

Because fidelity, signal-to-noise ratio, and distortion in any amplifier depend largely on the quality of that all-important first stage of amplification, this JFET-input preamp can improve the performance of most stereo sets. Or you can use it in front of a mixer to improve signal-to-noise ratio. Benefits include a very high input resistance so as not to degrade weak signals, a very low noise figure, and a virtually flat frequency response curve from 20 Hz. through 20 KHz. Also, continuous duration output short-circuit protection will guard against mistakes in interconnections. Adjustable voltage gain (unity to times 11) allows you to increase the signal strength of your commercial amplifier so that it can operate on a lower, more linear portion of its curve. It drives any amplifier with 2,000 ohms or more input resistance.

For construction, use dual complementary supply number 11-8 (set for ±12 volts), 11-9 (set for ±15 volts) or 11-16. To avoid hum pickup, position circuit components close together and away from those of the power supply (especially the transformers). Keep leads and interconnections as short as possible. Also, use shielded audio-type cable with the protective outer braid connected to circuit common for routing the input signals between the input jacks and the IC pins. Position "Gain Adjust" variable resistors R_3 and R_{3A} for front-panel access. To avoid damage to the ICs, ensure that they are correctly installed in their sockets and that the power supply is connected in the proper polarity before energizing this circuit.

For the initial adjustment, begin with both Gain Adjust variable resistors R_3 and R_{3A} set to their lowest values. Then

Inexpensive Semiconductors for Top-Quality Sound 127

increase each by a like amount until the best stereo sound results from your commercial amplifier. Note that too much gain from this preamp can cause distortion of high value signals (clipping). For optimal results, use an oscilloscope to observe waveforms while making this setting.

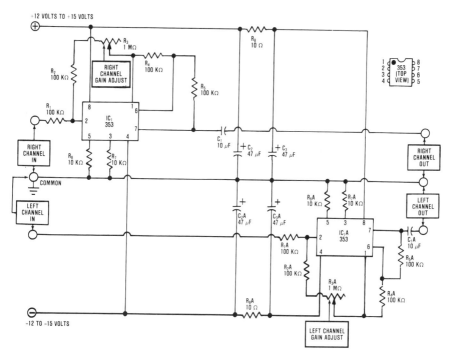

Circuit 6-6. Super stereo preamp

PARTS LIST

C_1, C_{1A}—10-μF, 25-volt, non-polarized electrolytic (Radio Shack 272-999)
C_2, C_{2A}, C_3, C_{3A}—47-μF, 25-volt aluminum electrolytic(Radio Shack 272-1015 or 272-1027)
IC_1, IC_{1A}—353 integrated circuit (Radio Shack 276-1715)
R_1, R_{1A}, R_2, R_{2A}, R_4, R_{4A}, R_5, R_{5A}—100,000 ohm, ¼-watt (Radio Shack 271-1347)
R_3, R_{3A}—1,000,000 ohm, ¼-watt (or more), linear taper potentiometer (Radio Shack 271-211)
R_6, R_{6A}, R_7, R_{7A}—10,000 ohm, ¼-watt (Radio Shack 271-1335)
R_8, R_{8A}—10 ohm, ¼-watt (Radio Shack 271-1301)

Circuit 6-7. Hi-Fi Magnetic Cartridge Preamp

Add the superior performance of a magnetic phonograph cartridge to a stereo amplifier having only ceramic phono or auxiliary inputs. Also ideal for adding a second magnetic cartridge-equipped turntable through an auxiliary input, or as a preamp for a high-impedance microphone. Adjustable voltage gain (from 10 to 100) allows you to set amplification for optimal sound. High performance features include 20 Hz. to 20K Hz. frequency response, very high input impedance, very low noise figure, and a total harmonic distortion figure of less than 0.02%. Also, continuous short-circuit protection for each output.

For construction, use dual complementary supply 11-8 (set for ±12 volts), or 11-9 (set for ±15 volts), or 11-16. To avoid pickup of hum, try to position circuit parts close together and away from those of the power supply. Keep all leads and interconnecting wires as short as possible. Use audio-type shielded cable for routing the input signals between the input jacks and the pins of the ICs (connect the outer braid to circuit common). To avoid damage to the IC's, be sure to insert them into their sockets right-side-up and insure that the power supply is connected in correct polarity. Position "Gain Adjust" variable resistors R_5 and R_{5A} for front panel access.

For the initial adjustment, begin with the Gain Adjust resistors R_5 and R_{5A} set for minimal gain. Then increase the gain of each by a like amount for best sound performance. Note that too much gain could cause clipping of some high-value signals. For optimal adjustment, observe waveforms with an oscilloscope as you make this adjustment.

PARTS LIST

C_1, C_{1A}—10-μF, 25-volt, non-polarized electrolytic (Radio Shack 272-999)
C_2, C_{2A}, C_3, C_{3A}—47-μF, 25-volt aluminum electrolytic (Radio Shack 272-1015 or 272-1027)
IC_1, IC_{1A}—353 integrated circuit (Radio Shack 276-1715)
R_1, R_{1A}, R_3, R_{3A}, R_4, R_{4A}—100,000 ohms, ¼-watt (Radio Shack 271-1347)
R_2, R_{2A}—1,000,000 ohms, ¼-watt (Radio Shack 271-1356)
R_5, R_{5A}—1,000,000 ohms, ¼-watt (or more), linear taper potentiometer (Radio Shack 271-211)
R_6, R_{6A}, R_7, R_{7A}—10,000 ohms, ¼-watt (Radio Shack 271-1335)
R_8, R_{8A}—10 ohms, ¼-watt (Radio Shack 271-1301)

Inexpensive Semiconductors for Top-Quality Sound 129

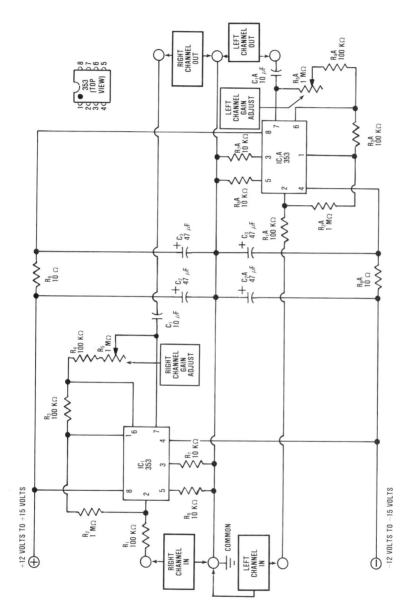

Circuit 6-7. Hi-Fi magnetic cartridge preamp

Circuit 6-8. Professional Stereo Mixer

Precisely blend your group's performance! Four independent inputs each accept high or low impedance microphone, phonograph, tape, or auxiliary signals (high level). Each input is continually adjustable from unity gain to -10 dB. Stereo/Mono switch allows you to separate inputs for stereo recordings, or to mix them together for mono. Performance features include full-spectrum sound response (20 Hz or 20K Hz.), very high input impedance, very low distortion, and complete output short-circuit protection. Drives any power amp with an input impedance of 2,000 ohms or more.

For construction tips, use dual complementary supply 11-8 (set for ±12 volts), or 11-9 (set for ±15 volts), or 11-16. Try to mount all circuit components close together and away from power supply components (especially the transformers). Of course, keep leads

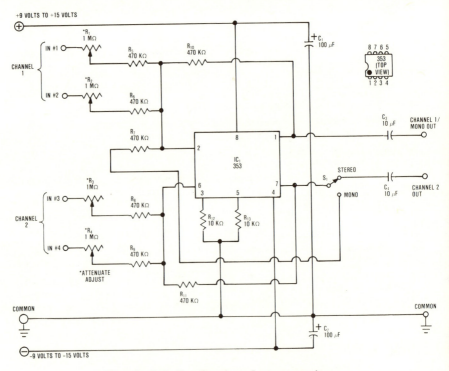

Circuit 6-8. Professional stereo mixer

Inexpensive Semiconductors for Top-Quality Sound 131

and interconnections as short as possible. Use audio-type shielded cable for routing the input signals between the input jacks and IC pins (connect the outer braid to circuit common). Avoid damage to the IC by insuring that it is inserted into its socket right-side-up, and that the power supply is connected in correct polarity. Position "Attenuate Adjust" R_1 through R_4 plus the Mono-Stereo switch for front panel access.

PARTS LIST

C_1, C_2—100-μF, 25-volt aluminum electrolytic (Radio Shack 272-1016 or 272-1028)
C_3, C_4—10-μF, 25-volt, non-polarized electrolytic (Radio Shack 272-999)
IC_1—353 integrated circuit (Radio Shack 276-1715)
R_1, R_2, R_3, R_4—1,000,000 ohm, ¼-watt (or more), audio taper potentiometer
R_5, R_6, R_7, R_8, R_9, R_{10}, R_{11}—470,000 ohm, 5% tolerance, ¼-watt (Radio Shack 271-1354)
R_{12}, R_{13}—10,000 ohm, ¼-watt (Radio Shack 271-1335)
S_1—SPDT switch (Radio Shack 275-613)

In closing, the complex and bulky audio amplifiers of the past are largely obsolete in many cases because tiny IC chips can perform the same function at a fraction of the cost. And to the forward-looking technician, these developments can mean real opportunities for fun and profit in this fascinating and explosively growing field of electronics.

7

Practical Semiconductor Circuits for Your Car

The miniature size and low energy consumption of microelectronics makes them a natural for use in your car, truck, or van. In ways only imagined a short while ago, you can use them to enhance the safety and convenience of your vehicle... at a small fraction of the dealer price. This chapter contains a variety of tested formulas to show you how.

Fortunately, the major problem that plagued early automotive electronic systems is now largely solved. As service mechanics recall all too well, even the simple act of disconnecting a battery cable could produce a voltage transient capable of harming delicate semiconductor components. (transients are regularly generated in automotive electric systems for a variety of reasons). However, by installing Decoupler/Regulator Circuit 11-21 where it is called for in this chapter, you can protect circuits from damage. Here are a few examples to give you an overview of this chapter.

Insurance company statistics indicate that many accidents occur as vehicles with limited rear vision are backed up. If you drive a van, truck, or camper, consider Back-up Sounder Circuit 7-1. It will sound a pulsating tone as you shift your vehicle into reverse, thereby alerting people and pets in the immediate vicinity.

As another example, cold-weather driving conditions can place a severe strain on your car's battery—actually reducing its efficiency by half. This means that even if your battery is in good condition, it may not have the power to start your engine in the morning if—for example—you drove in traffic with your head lamps ON the night before. If this is the problem, battery charger Circuit 7-3 can be the answer. It will not only top off your battery with energy during difficult driving conditions, but you can also use it for recharging a completely dead battery or for trickle-charging purposes.

You'll also find a variety of circuits which can help you avoid a loss. For example, Circuit 7-4 will warn you with a buzz should your brake fluid run low. This is important because your car would stop poorly, if at all, should this vital fluid run out. And because it is so easy to overlook an "idiot light" which has come ON indicating trouble, Circuit 7-5 can keep tabs on the situation and let you know with a buzz should trouble occur. As a few more examples, Circuit 7-6 will inform you in the event that you accidently leave your headlamps ON, and intrusion alarm Circuit 7-7 will keep a watch on your car while you're away, activating a loud sounder should anyone attempt to enter.

Circuit 7-1. Back-up Sounder

Avoid trouble before it occurs with this warning device. Connects to the back-up lamp of any standard 12-volt, negative ground system, and sounds a pulsating warning tone whenever the vehicle is shifted into reverse. Sufficiently loud to alert pets, children, and older folks in the immediate vicinity, it is a should-have item for vehicles with limited rear vision such as vans and recreational vehicles.

For construction, use decoupler/regulator circuit 11-21 connected to a source which is hot while the ignition switch is ON. Mount piezo buzzer at the rear of the vehicle in a location which is protected from the elements. For making the connection to the back-up lamp, consult your vehicle wiring diagram for the most convenient tap-in point (as shown in the schematic). With the ignition switch ON and the vehicle in reverse, rotate "Tone Adjust" R_4 for the desired sound.

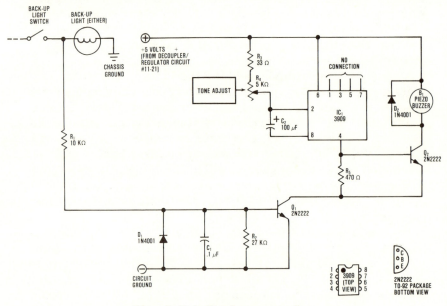

Circuit 7-1. Back-up sounder

PARTS LIST

B_1—5-volt piezo buzzer (Radio Shack 273-060)
C_1—0.1—μF, 50-volt ceramic disk (Radio Shack 272-135)
C_2—100-μF, 16-volt aluminum electrolytic (Radio Shack 272-1016 or 272-1028)

Practical Semiconductor Circuits for Your Car **137**

D_1, D_2—1N4001 diodes (Radio Shack 276-1101)
IC_1—3909 integrated circuit (Radio Shack 276-1705)
Q_1, Q_2—2N2222 or similar transistors (Radio Shack 276-1617)
R_1—10,000 ohm, ¼-watt (Radio Shack 271-1335)
R_2—27,000 ohm, ¼-watt (Radio Shack 271-1340)
R_3—33 ohm, ¼-watt (Radio Shack 271-007)
R_4—5,000 ohm, ¼-watt (or more), linear taper potentiometer (Radio Shack 271-1714)
R_5—470 ohm, ¼-watt (Radio Shack 271-1317)

Circuit 7-2. Handy 12-Volt to 120-Volt Inverter

Imagine a low-power electrical outlet in your auto or van! A capacity of better than 15 watts allows you to operate small fluorescent lamps, shavers—almost any kind of appliance which doesn't synchronize with line frequency. Home-type stereo units are OK as long as you keep the volume low. Circuit automatically shuts itself down if the load is too great. And with only nine parts, it's extraordinarily easy to construct.

When making this circuit, remember that high voltage will be present, and act accordingly. House-type electrical hardware is recommended for the high-voltage portion of this circuit.

For driving audio equipment, a somewhat larger value of capacitor for C_1, may be substituted in order to reduce hum. A certain amount of hum, however, is normal because this circuit produces a square-wave output.

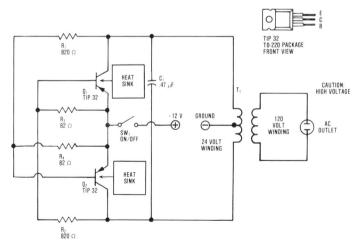

Circuit 7-2. Handy 12-volt to 120-volt inverter

PARTS LIST

C_1—0.47-μF, 100-volt, mylar (Radio Shack 272-1054)
Q_1, Q_2—TIP32 or similar transistors (Radio Shack 276-2025)
R_1, R_2—820 ohm, 1/2-watt (Shunt Two Radio Shack 271-026)
R_3, R_4—82 ohm, 1/2-watt (Radio Shack 271-011)
SW_1—2-amp on/off switch (Radio Shack 275-602)
T_1—24-volt, 2-amp, center-tapped secondary, heavy-duty transformer (Radio Shack 273-1512 is OK)

Circuit 7-3. 12-Volt Automotive Battery Charger

Top off your battery with energy during severe winter driving conditions ... Bring a discharged battery up to specifications (24 to 36 hours)... Or trickle-charge your battery for improved service and longer life. This charging circuit outputs up to 1½-amps, and features automatic rate reduction as the battery comes up to charge. A convenient test point allows you to directly read rate of charge using any standard voltmeter.

For construction of this circuit, an IC regulator housed in a TO-3 package is recommended for highest continuous output current. But in any event, be sure to heat-sink the regulator very well. Position "Charge Level" potentiometer R_3 and "Test Point 1 and 2" for easy access from the front of the circuit box. Use heavy jumper clips for making connection with the battery.

For the initial circuit adjustment, attach a dc voltmeter set to easily read 1½ volts across test points 1 and 2, and connect this circuit across a fully charged 12-volt battery in good condition (engine off). Next, set Charge Level R_3 for a trickle rate of about 25 to 40 mA (.025 to .040 volts).

Then for service, simply connect this circuit across the battery to be charged. It will automatically adjust its current rate according to the charge level of the battery, tapering off to a trickle as the battery comes up to full.

PARTS LIST

C_1—1,000-μF, 35-volt aluminum electrolytic (Radio Shack 272-1019 or 272-1032)
C_2, C_3—100-μF, 25-volt aluminum electrolytic (Radio Shack 272-1016 or 272-1028)
D_1, D_2—1N4002 diodes (Radio Shack 276-1102)
IC_1—7805 or LM340-5.0 3-terminal voltage regulator (Radio Shack 276-1770)

Practical Semiconductor Circuits for Your Car   139

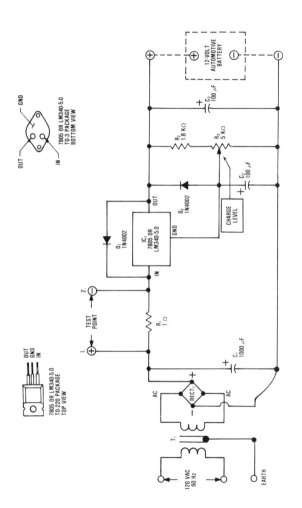

Circuit 7-3. 12-volt automotive battery charger

R_1—1 ohm, 2-watt (or more) (Radio Shack 271-131)
R_2—1,800 ohm, ¼-watt (Radio Shack 271-1324)
R_3—5,000 ohm, ¼-watt (or more), linear taper potentiometer (Radio Shack 271-1714)
$Rect_1$—4-amp, 50 PIV, modular bridge rectifier (Radio Shack 276-1146)
T_1—18 volt, 2-amp secondary, heavy-duty transformer (Radio Shack 273-1515)

Circuit 7-4. Low Brake Fluid Monitor

Because a vehicle would stop poorly if at all should the brake fluid leak out, it's important to keep a close watch on the level of this vital liquid. Here's a circuit which will continuously "feel" for the presence of brake fluid in the master cylinder and warn you with an audio tone and a glowing LED should trouble arise. Switch S_1 lets you silence the sounder, although LED will continue to glow as a constant reminder. Ideal for use in a central control panel for monitoring your vehicle's operating status at a glance.

For construction advice, only top-quality construction and installation practices are acceptable with this safety unit. For power, use decoupler/regulator circuit 11-21 attached to a source

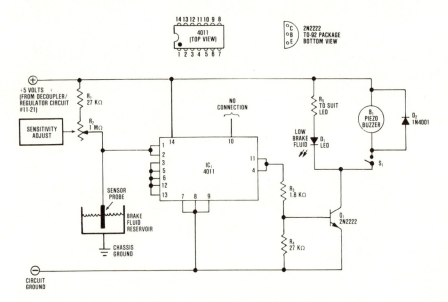

Circuit 7-4. Low brake fluid monitor

that is hot while the ignition switch is ON. Almost any metallic object will do for a sensor probe, although a length of bare, 20-gauge copper wire works well. Carefully tap a hole in the master cylinder cap and insert this probe into the reservoir so that it will remain immersed in any normal level of fluid. For a metallic cap which grounds to the master cylinder case, you must insulate the probe-to-cap junction. For a master cylinder with dual reservoirs, either reservoir may be monitored, although the one supplying fluid to the rear brakes is the more important. At any rate, insure that the hole you tap is above the fluid level to prevent leaks, and completely air-tight to prevent the entrance of moisture-laden air (brake fluid readily absorbs moisture).

Finishing up, set the "Sensitivity Adjust" variable. Resistor R_2 to its mid position. In some rare cases it may be necessary to increase/decrease sensitivity from this point. Remember that this circuit is only a back-up system and that the fluid level must still be periodically checked.

PARTS LIST

B_1—Any 5-volt, low-current piezo buzzer (Radio Shack 273-060)
D_1—Any red light-emitting diode (such as Radio Shack 276-026)
D_2—1N4001 diode (Radio Shack 276-1101)
IC_1—4011 integrated circuit (Radio Shack 276-2411)
Q_1—2N2222 or similar transistor (Radio Shack 276-1617)
R_1, R_4—27,000 ohm, ¼-watt (Radio Shack 271-1340)
R_2—1,000,000 ohm, ¼-watt, linear taper potentiometer (Radio Shack 271-229)
R_3—1,800 ohm, ¼-watt (Radio Shack 271-1324)
R_5—To suit LED
S_1—ON/OFF switch (Radio Shack 275-602)

Circuit 7-5. Idiot Light Audio Back-Up

It can be a costly matter if you fail to notice that one of the "idiot lights" on your dashboard has come ON indicating trouble. Give your "oil" and "temperature" warning lamps an audio back-up with this complete system. A 12-second delay mechanism permits normal starting of the engine without activating sounder. Switch S_1 allows you to silence the sounder although LED D_1 continues to glow as a constant reminder. Ideal for use in a central control panel for monitoring your vehicle's operating status at a glance. Fits any standard 12-volt negative ground system.

For assembly and installation, power this circuit with decoupler/regulator circuit 11-21 attached to a source that is hot

142 *Practical Semiconductor Circuits for Your Car*

while the ignition switch is ON. The sensor wires leading to the oil and temperature lamps may be connected in either order (consult your vehicle's wiring diagram for the most convenient tap-in points). To test circuit operation, connect a dc voltmeter to Test Point 1 as shown on the schematic diagram. Then turn ON the ignition switch without starting the engine so that the warning lamps glow. The voltage seen at Test Point 1 will slowly rise, triggering the sounder after about a 12-second interval. In service, either lamp coming ON is sufficient to activate the sounder.

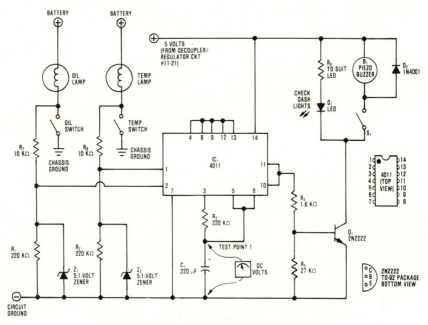

Circuit 7-5. Idiot light audio back-up

PARTS LIST

B_1—Any 5-volt, low current piezo buzzer (Radio Shack 273-060)
C_1—220-μF, 16-volt aluminum electrolytic (Radio Shack 272-1006 or 272-956)
D_1—Any red light-emitting diode (Radio Shack 276-026)
D_2—1N4001 diode (Radio Shack 276-1101)
IC_1—4011 integrated circuit (Radio Shack 276-2411)
Q_1—2N2222 or similar transistor (Radio Shack 276-1617)
R_1, R_2, R_3—220,000 ohm, ¼-watt (Radio Shack 271-1350)
R_4—1,800 ohm, ¼-watt (Radio Shack 271-1324)
R_5—27,000 ohm, ¼-watt (Radio Shack 271-1340)
R_6—To suit LED
R_7, R_8—10,000 ohm, ¼-watt (Radio Shack 271-1335)
S_1—ON/OFF switch (Radio Shack 275-602)
Z_1, Z_2—5.1 volt, 1-watt zener diodes (1N4733)

Circuit 7-6. Headlamp-ON Reminder

Never again will you return to your car to find your battery run down because you accidently left the headlamps ON. This "smart" circuit constantly monitors the voltage levels in your vehicle's electrical system, and sounds a tone when the headlamps are ON but the ignition switch is OFF. A 5-sound delay mechanism permits normal operation of controls yet triggers the sounder before you can exit from the vehicle. Connects to any standard 12-volt system. Ideal for use in a central control panel for monitoring your vehicle's operating status at a glance.

For construction pointers, power this circuit with decoupler/regulator circuit 11-21 connected to a source that is hot, regardless of the position of the ignition switch. The sensor wires leading to an ignition-switch-activated fuse and to a headlamp-switch-activated point in the wiring system are not interchangeable and must be connected as shown. Consult your vehicle's wiring diagram for the most convenient tap-in points.

PARTS LIST

B_1—Any 5-volt, low-current piezo buzzer (Radio Shack 273-060)
C_1—1-μF, 25-volt solid tantalum (Radio Shack 272-1419)
C_2—0.1-μF, 50-volt ceramic disk (Radio Shack 272-135)
C_3—100-μF, 25-volt aluminum electrolytic (Radio Shack 272-1016 or 272-1028)
D_1—1N4001 diode (Radio Shack 276-1101)
D_2—Any red light-emitting diode (such as Radio Shack 276-026)
IC_1—4011 integrated circuit (Radio Shack 276-2411)
IC_2—555 integrated circuit (Radio Shack 276-1723)
Q_1—2N3906 or similar transistor (Radio Shack 276-2034)
R_1, R_2—10,000 ohm, ¼-watt (Radio Shack 271-1335)
R_3, R_4—220,000 ohm, ¼-watt (Radio Shack 271-1350)
R_5—1,000 ohm, ¼-watt (Radio Shack 271-1321)
R_6—27,000 ohm, ¼-watt (Radio Shack 271-1340)
R_7—10,000 ohm, ¼-watt (Radio Shack 271-1335)
R_8—56,000 ohm, ¼-watt (Radio Shack 271-043)
R_9—To suit LED
Z_1, Z_2—5.1-volt, 1-watt zener diodes (1N4733)

144 *Practical Semiconductor Circuits for Your Car*

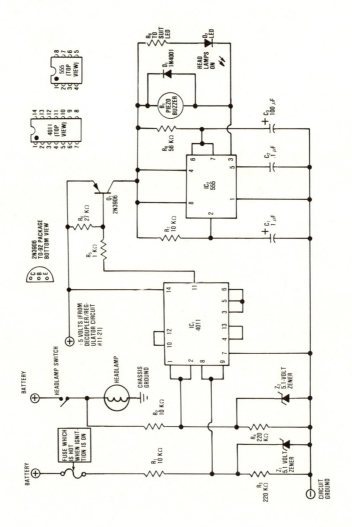

Circuit 7-6. Headlamp-ON reminder

Circuit 7-7. A Dependable Intrusion Alarm for Your Car, Truck or Van

Protect your vehicle and the valuables inside against theft with this complete security system. Simply activate this alarm by turning key in lock switch S_1. Then if vandals try to open door, trunk, hood—anywhere that you have installed a push-button switch—a loud sounder is activated for about a one-minute interval. (To increase this time interval, use a larger-value capacitor for C_2.) An LED glows to inform you if the alarm has been triggered while you were away. Alarm is easily reset by momentarily opening lock switch S_1. Suitable for any standard 12-volt negative ground system.

For assembly and installation, use decoupler/regulator circuit 11-21 attached to a source that is always hot to power this circuit. Mount lock switch S_1 in a convenient exterior location such as under the driver's door key lock. Switches $S_2, S_3 \ldots S_x$ are

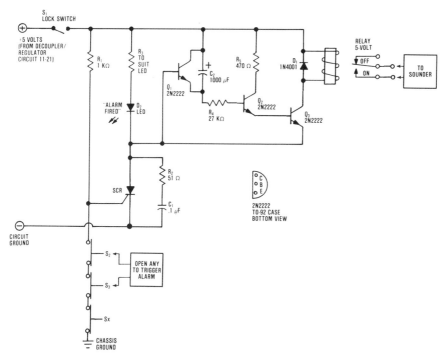

Circuit 7-7. A dependable intrusion alarm for your car, truck or van

normally open momentary-contact push-button switches installed on entry ports so that they are usually held closed (opening any trigger's alarm). Use the relay contacts to activate sounder which is wired directly into the vehicle's 12-volt system. Almost any loud bell, siren, or buzzer will do as long as it does not exceed the current rating of the relay contacts. For higher current demands, use a relay with multiple contacts shunted together. As a money-saving option, the relay contacts may be used to activate the horn and/or headlamp relays in a vehicle where these units are always connected to power (consult your vehicle's wiring diagram for specifics).

PARTS LIST

C_1—0.1-μF, 50-volt ceramic disk (Radio Shack 272-135)
C_2—1,000-μF, 16-volt aluminum electrolytic (Radio Shack 272-958)
D_1—1N4001 diode (Radio Shack 276-1101)
D_2—Any red light-emitting diode (such as Radio Shack 276-026)
Q_1, Q_2, Q_3—2N2222 or similar transistors (Radio Shack 276-1617)
R_1—1,000 ohms, ¼-watt (Radio Shack 271-1321)
R_2—51 ohm, ¼-watt (Radio Shack 271-1307 is OK)
R_3—To suit LED
R_4—27,000 ohm, ¼-watt (Radio Shack 271-1340)
R_5—470 ohm, ¼-watt (Radio Shack 271-1317)
Relay—Any 5 volt relay, 50-ohm or greater coil resistance (Radio Shack 275-215 or 275-216)
S_1—Lock switch (Radio Shack 49-515)
S_2, S_3—S_x—Momentary contact switches for doors, trunk, etc. (Radio Shack 49-513 or 275-1547)
SCR—Any medium-current silicon-controlled rectifier (Radio Shack 276-1067)

As a few parting comments, semiconductors can hardly be used to better advantage than in circuits for your car. We feel confident that any circuit described in this chapter will reward you for your construction efforts several times over through useful service and increased peace of mind.

Semiconductor Projects for "Fun" and Profit

Solid-state electronics doesn't, of course, have to be oriented toward all work and no play. In many cases this technology can help make your leisure-time activities more enjoyable—and profitable.

One prime example of this is CW oscillator circuit 8-1. As an aid in practicing your Morse Code for taking the FCC exam, it generates a pleasant audio tone each time you close the telegraph key contacts. And to enhance your learning experience even further, tone and volume are continuously adjustable to suit your individual preference.

Another fun, yet practical, idea is Magnetizer/De-Magnetizer Circuit 8-2. You can use it to create no-cost presents for the kids or for hobbies ... and then put it to work around your shop for such purposes as charging screwdrivers and tack hammers for increased function. Another use is to neutralize the unwanted charge in tape recorder heads, thereby reducing audio noise and restoring high-frequency performance. As an added plus, you can probably assemble this circuit entirely from scrap parts found in your electrical spares box.

And consider the Circuit 8-4 Electroplating Machine. It can put a shiny coating on small metallic parts to give them eye appeal and to protect them from harm (such as automotive and boating parts). Almost any public library can supply you with detailed plating instructions for electro-chemically applying a variety of metallic coatings—from zinc to gold and silver.

For amusement at parties and the like, try the Circuit 8-5 Lie Detector. You simply touch the input probes to the fingertip or palm of the subject's hand, and observe the resultant waveforms on your oscilloscope as you ask the subject any question that you want. And because it consists of only six components, you can snap it together in short order.

One more circuit which can lead a dual life as a toy and as a tool is the Sound Microscope of Circuit 8-7. You can use it to pick up previously inaudible vibrations such as those made by a distant train along the railroad tracks, or an insect chewing on a blade of grass. Alternately, it can earn its keep in a variety of ways such as by allowing you to listen for defects in delicate mechanisms, for example, a fishing reel.

Circuit 8-1. CW Oscillator For Morse Code Practice

Here's an easy-to-build circuit which can help you increase your Morse Code speed for taking the FCC exam. Quality features include tone control and volume control. And you may listen privately with headphones by simply including a jack. Operates on a single 9-volt battery or on a line-powered supply—as you prefer.

For construction pointers, use battery supply number 11-18 for portable operation, or line-powered supply number 11-4, 11-11,

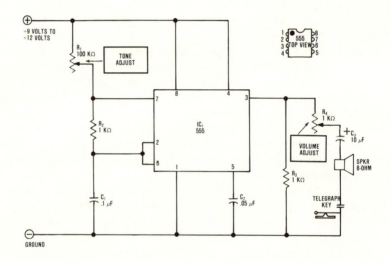

Circuit 8-1. CW oscillator for Morse Code practice

PARTS LIST

C_1—0.1-μF, 50-volt ceramic disk (Radio Shack 272-135)
C_2—0.05-μF, 50-volt ceramic disk (Radio Shack 272-134)
C_3—10-μF, 25-volt, aluminum electrolytic (Radio Shack 272-1013 or 272-1025)
IC_1—555 integrated circuit (Radio Shack 276-1723)
R_1—100,000 ohm, ¼-watt (or more), linear taper potentiometer (Radio Shack 271-092)
R_2, R_3—1,000 ohm, ¼-watt (Radio Shack 271-1321)
R_4—1,000 ohm, ¼-watt (or more), linear taper potentiometer
SPKR—Any 8-ohm speaker
Telegraph Key—Any telegraph key (such as Radio Shack 20-1084)

or 11-12 to save on battery replacement. Be sure to position "Tone Adjust" variable resistor R_1 and "Volume Adjust" variable resistor R_4 for easy access from the front of the circuit box. Initially, set both of these controls to their mid-positions, and then adjust from here. You may locate the speaker inside or outside the circuit box, whichever is convenient.

Circuit 8-2. Handy Magnetizer/De-Magnetizer

A highly useful project that you can build entirely from scrap parts. With dual-throw switch S_1 in the "Magnetize" position you can use this circuit for such purposes as charging screwdrivers and tack hammers to improve their functionality... or to make low-cost magnets as presents for the kids. With switch S_1 in the "De-Magnetize" position, you can remove unwanted charge from such objects as tape recorder heads.

For construction of this circuit, first secure a 12-volt winding rated for continuous duty such as that from a scrap solenoid, relay, etc. The central opening of this winding should be wide enough for insertion of parts such as screwdrivers and tape recorder heads for magnetic interaction. Next, measure the resistance of this winding with an ohmmeter, and calculate the current which the transformer must deliver using this formula:

$$\text{Current} = \frac{12 \text{ volts}}{\text{coil resistance of winding in ohms}}$$

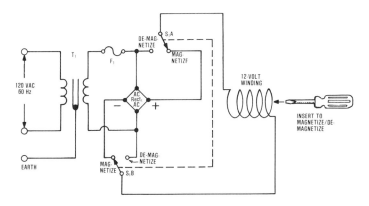

Circuit 8-2. Handy magnetizer/de-magnetizer

Finishing up, select a 12-volt transformer capable of delivering twice this amount of current ... a fuse to match the transformer's current rating ... and a modular bridge rectifier of double the transformer's current rating (and of 50 PIV). Position magnetize/de-magnetize switch S, for easy access from the front of the circuit box.

PARTS LIST

F_1—Fuse to match transformer's current rating
$Rect_1$—Modular bridge rectifier of 50 PIV and of double the transformer's current rating
S_{1A}, S_{1B}, Dual-pole, dual-throw switch (Radio Shack 275-666)
T_1—12-volt or 12.6-volt transformer of appropriate current rating
12-Volt Winding—Winding from a scrap 12-volt solenoid, relay, etc.

Circuit 8-3. Beat Generator

You'll wonder how you ever got along without it! Generates an audio pulse and visual strobe at precisely adjustable intervals of about 20 beats per second through 1 beat every 2½ minutes. Hundreds of practical applications at home or work, including use as a metronome to coordinate your musical group's activities, or to

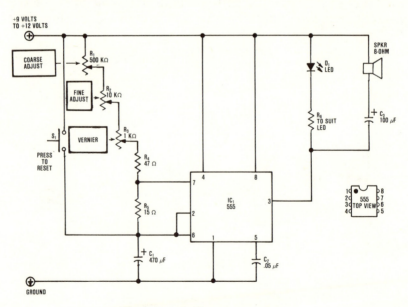

Circuit 8-3. Beat generator

Semiconductor Projects for "Fun" and Profit 153

call cadence for physical exercises. Can be worth its weight in gold for such purposes as developing photographic film where you must shake the developing canister at precise, 30-second intervals.

For assembly, you can use either battery supply number 11-18 for portable operation or plug-in supply number 11-4, 11-11, or 11-12 for service at a fixed location. Mount "Coarse Adjust" variable resistor R_1, "Fine Adjust" variable resistor R_2, "Vernier" variable resistor R_3, and "Press To Reset" switch S_1 for front-panel access.

In operation, "Coarse Adjust" varies the time interval between beats by minutes ... "Fine Adjust" by seconds ... and "Vernier" by fractions of a second. Pushing "Press To Reset" switch S_1 for about a second starts the count at zero during long-time interval use. Note that when this circuit is first energized, it takes a few seconds to begin operation.

PARTS LIST

C_1—470-μF, 25-volt aluminum electrolytic (Radio Shack 272-1018 or 272-1030)
C_2—0.05-μF, 50-volt ceramic disk (Radio Shack 272-134)
C_3—100-μF, 25-volt aluminum electrolytic (Radio Shack 272-1016 or 272-1028)
D_1—Any red, light-emitting diode (optional) such as Radio Shack 276-026)
IC_1—555 integrated circuit (Radio Shack 276-1723)
R_1—500,000 ohm, ¼-watt (or more), linear taper potentiometer (Radio Shack 271-210)
R_2—10,000 ohm, ¼-watt (or more), linear taper potentiometer (Radio Shack 271-1715)
R_3—1,000 ohm, ¼-watt or more, linear taper potentiometer
R_4—47 ohm, ¼-watt (Radio Shack 271-1307)
R_5—15 ohm, ¼-watt (Radio Shack 271-003)
R_6—To suit LED
S_1—Normally open momentary contact switch (optional) (Radio Shack 275-1547)
SPKR—Any 8-ohm speaker

Circuit 8-4. Electroplating Machine

Give small metallic parts a shining coat to enhance their appearance, to protect them from corrosion, or to improve their electrical conductivity. Coatings of copper, zinc, tin, chromium, nickel—even gold and silver—are possible (consult your public library for detailed plating instructions). Makes use of your bench-type dc ampmeter to keep construction costs low.

For circuit construction, be sure to heat-sink the IC regulator well so that it will deliver full current output. In addition, a regulator housed in a TO-3 package is recommended. Position "Current Adjust" potentiometer R_2 and "Point 1 and Point 2" for easy access from the front of the circuit box.

Here are some simplified plating instructions to get you started: Dilute two tablespoons of copper sulfate ($CuSO_4$—available in many drug and hardware stores) into a pint of water in a plastic tub. Thoroughly clean the object to be plated to bare, clean metal with a steel-wool pad. Next, attach a wire to the

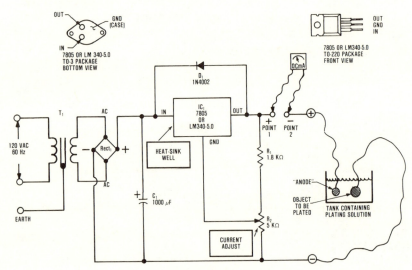

Circuit 8-4. Electroplating machine

PARTS LIST

C_1—1,000-μF, 25-volt aluminum electrolytic (Radio Shack 272-1019 or 272-1032)
D_1—1N4002 diode (Radio Shack 276-1102)
IC_1—7805 or LM340-5.0 3-terminal voltage regulator (Radio Shack 276-1770)
R_1—1,800 ohm, ¼-watt (Radio Shack 271-1324)
R_2—5,000 ohm, ¼-watt (or more), linear taper potentiometer (Radio Shack 271-1714)
$Rect_1$—2-amp, 50 PIV, modular bridge rectifier (Radio Shack 276-1146)
T_1—12-volt or 12.6-volt, 1-amp secondary, standard transformer (Radio Shack 273-1505)

Semiconductor Projects for "Fun" and Profit   155

positive output terminal ... spot-solder this wire to a copper penny ... and immerse the penny in the solution (this is to avoid metallic contamination as the anode dissolves). Then attach a wire to the negative output terminal ... spot-solder this wire to the object to be plated ... and immerse the object in the solution.

For electroplating, energize the circuit and set Current Adjust R_2 to about 100 mA for a quarter dollar-sized object, larger objects proportionally more. Ventilate well. Note that too much current will result in the formation of sludge. Shift object occasionally for a more even coating. Somewhat more current may be necessary as the plate thickness increases. When plating is complete, buff object to a high gloss.

Circuit 8-5. Lie Detector

Great for party games. Works on the principle that the electrical conductivity of skin changes with an emotional state. Simply touch input probes to fingertip or palm of subject's hand, and observe resultant waveforms on your oscilloscope. Square-wave pattern will increase in frequency when subject tells a lie or otherwise becomes nervous.

For construction tips, use battery supply 11-18 for portable operation, or plug-in supply 11-1, 11-4, 11-10, 11-11, or 11-12 to

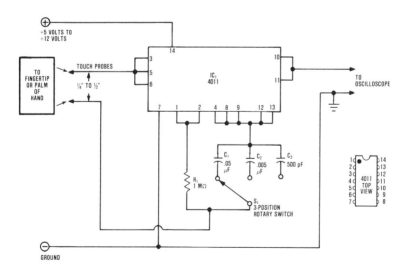

Circuit 8-5. Lie detector

eliminate battery replacement. Mount rotary switch S_1 for front-panel access. Affix touch probes ¼ to ½-inch apart on a rigid structure.

To operate this circuit, contact subject's palm or fingertip with touch probes, and turn rotary switch S_1 for best square-wave pattern on your oscilloscope. Next, ask subject neutral questions (such as Social Security number and the answer to multiplication tables) for several minutes until some ambient waveform frequency is established. Then when you ask more pointed questions, the waveform will usually increase in frequency if a truthful answer is not forthcoming.

PARTS LIST

C_1—0.05-μF, 50-volt ceramic disk (Radio Shack 272-134)
C_2—0.005-μF, 50-volt ceramic disk (Radio Shack 272-130)
C_3—500 PF, 50-volt ceramic disk (Radio Shack 272-125)
IC_1—4011 integrated circuit (Radio Shack 276-2411)
R_1—1,000,000 ohm, ¼-watt (Radio Shack 271-1356)
S_1—3-position rotary switch (Radio Shack 275-1385 is OK)

Circuit 8-6. Sports Starting Light Sequencer

Get sport races off to a smooth and even start with this 1-2-3-GO light sequencer ("Christmas tree" lights). Continuously adjustable countdown rate allows optimal starting of a wide variety of events. Drives flashlight-type lamps/low current buzzers directly, or more powerful lamps/buzzers through relays. May be energized either by batteries or by a line-powered supply.

For assembly, use battery supply number 11-20 (or connect directly to the poles of a 12-volt battery in an automobile) ... or line-powered supply 11-4, 11-5, 11-7, or 11-12 (depending on your current requirements). Remember that output devices (lamps, buzzers, etc.) are most effective when mounted in a fixture positioned at or just above the eye level of the contestants. Each output driver-transistor (as exemplified by Q_2 in the schematic) will drive a flashlight-type lamp (such as a Radio Shack 272-1143) or a low-current buzzer directly ... or may be used to close a 12-volt relay for switching ON a more powerful device. Be sure to mount "Rate Adjust" potentiometer R_1 and "Standby/Countdown" switch S_1 for access from the front of the circuit box.

Semiconductor Projects for "Fun" and Profit 157

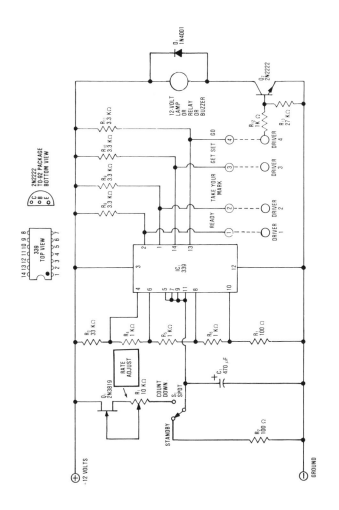

Circuit 8-6. Sports starting light sequencer

To operate this circuit, simply click switch S_1 to the Countdown position in order to cycle-ON the starting lights. The rate at which the lights cycle-ON, which must be preset before the race, is adjusted by rotating Rate Adjust R_1. To start the next race, simply return switch S_1 to the Standby position and the circuit is ready to begin the next countdown.

PARTS LIST

C_1—470-μF, 25-volt, aluminum electrolytic (Radio Shack 272-1018 or 272-1030)
D_1—1N4001 diode (Radio Shack 276-1101)
IC_1—339 integrated circuit (Radio Shack 276-1712)
Q_1—2N3819 or similar N-channel FET (Radio Shack 276-2035)
Q_2—2N2222 or similar transistor (Radio Shack 276-1617)
R_1—10,000 ohm, ¼-watt (or more), linear taper potentiometer (Radio Shack 271-1715)
R_2, R_7—100 ohm, ¼-watt (Radio Shack 271-1311)
R_3—33,000 ohm, ¼-watt (Radio Shack 271-1341)
R_4, R_5, R_6, R_{12}—1,000 ohm, ¼-watt (Radio Shack 271-1321)
R_8, R_9, R_{10}, R_{11}—3,300 ohm, ¼-watt (Radio Shack 271-1328)
R_{13}—27,000 ohm, ¼-watt (Radio Shack 271-1340)
S_1—Single pole, dual-throw switch (Radio Shack 275-613)

Circuit 8-7. Sound Microscope

Explore the world of sound vibrations inaudible to the unaided ear! Just a few of the fun and practical applications for this circuit include listening to the footsteps of an insect, or evaluating the condition of bearings inside a delicate mechanism such as a fishing reel. Operates on batteries for portability and reduction of 60-cycle hum. May be used with headphones simply by installing a jack.

For construction, power this circuit with a combination of dual-complementary supply number 11-19 plus a 6-volt or 12-volt lantern-type battery. For the pickup device, use an old ceramic phonograph pickup cartridge (the needle is unnecessary for most applications). With a stereo-type cartridge, you may connect to either channel. If you don't have a ceramic cartridge available, a crystal microphone element will usually work. Mount "Gain Adjust" Variable Resistor R_5 and "Volume Adjust" potentiometer R_8 for easy access from the front of the circuit box.

To operate this circuit, position the phonograph cartridge pickup device for maximum coupling with the sound vibration that you wish to hear. Taping the cartridge in direct physical

Semiconductor Projects for "Fun" and Profit 159

contact with the vibrating medium is often the most practical way to accomplish this. Then rotate Gain Adjust R_5 and Volume Adjust R_8 for best sound performance.

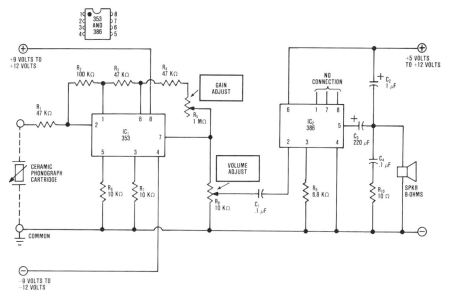

Circuit 8-7. Sound microscope

PARTS LIST

C_1, C_4—0.1-μF, 50-volt ceramic disk (Radio Shack 272-135)
C_2—1-μF, 25-volt solid tantalum (Radio Shack 272-1419)
C_3—220-μF, 25-volt aluminum electrolytic (Radio Shack 272-1017 or 272-1029)
IC_1—353 integrated circuit (Radio Shack 276-1715)
IC_2—386 integrated circuit (Radio Shack 276-1731)
R_1, R_3, R_4—47,000 ohm, ¼-watt (Radio Shack 271-1342)
R_2—100,000 ohm, ¼-watt (Radio Shack 271-1347)
R_5—1,000,000 ohm, ¼-watt (Radio Shack 271-1356)
R_6, R_7—10,000 ohm, ¼-watt (Radio Shack 271-1335)
R_8—10,000 ohm, ¼-watt (or more), linear taper potentiometer (Radio Shack 271-1715)
R_9—6,800 ohm, ¼-watt (Radio Shack 271-1333)
R_{10}—10 ohm, ½-watt (Radio Shack 271-001)
SPKR—Any 8-ohm speaker

To cap off this chapter, the thing to remember with semiconductors is that they can serve you at play as well as at work. This can mean more enjoyment during your leisure hours... and new opportunities for profit!

Dynamic Semiconductor Warning Mechanisms

Of all the possible applications for semiconductors, one of the most beneficial is their use in alarm circuits to warn of impending trouble. In effect, you can let these circuits do your "worrying" for you. But since reliability in this application is more important than ever, we'll first discuss some ideas that will help you get the most out of your circuits. Then we'll highlight the useful nature of the units contained in this chapter.

On the topic of reliability, one prime requirement of an alarm system is that it must attract attention once it has been triggered. Common sense is your best guide here. For an abnormal-condition alarm, this can mean selecting a sounder that is loud enough to be heard from anywhere that you might be in your home. In the case of a burglar alarm, this usually means mounting the sounder so that it can be easily heard from outside your home, and then asking neighbors to phone the police in the event that they hear a one-minute blast. Each intrusion-type alarm in this chapter is equipped with a one-second adjustment for testing purposes ... a one-minute adjustment for normal service.

Yet again, reliability also hinges on the alarm circuitry receiving a continuous supply of power ... and being built to withstand ambient temperature conditions. For a line-powered alarm, you can practically guarantee uninterrupted service by using the battery back-up unit illustrated in Figure 9-1. This back-up unit is of particular advantage for a home-type intrusion alarm as some burglars always sever the power lines to a dwelling before attempting to enter. And if you plan to use one of these alarms in temperature extremes, you'll of course want to build it with components of the appropriate ratings.

BUILDING SILENT SENTRIES

At many locations around your home or workplace where mishaps are most likely to occur, a semiconductor alarm mechanism could be standing silent vigil for you. For example, if there's a small child in your family, Circuit 9-1 could be watching over your medicine cabinet. It senses when the cabinet door is opened, and then activates a loud buzzer if this door isn't closed after a preset time interval.

164   *Dynamic Semiconductor Warning Mechanisms*

Other alarms will constantly "feel" for the presence of high water/high temperature/low temperature (Circuits 9-2, 9-3, and 9-4 respectively). And should any of these conditions occur, the circuit will activate a sounder in order to give you early warning.

In a very real way, Breakbeam Alarm Circuit 9-5 can function as a second pair of "eyes" for you. When a beam of light striking its photocell pickup device is interrupted by someone stepping between light and cell, it beeps an audio notice. You can use it to keep track of the comings and goings of friends and customers, and even for security purposes.

And finally, rather than worry about thefts, consider the schematics of Circuits 9-6, 9-7, and 9-8 instead. They will show you innovative ways to enhance your feelings of security—electronically.

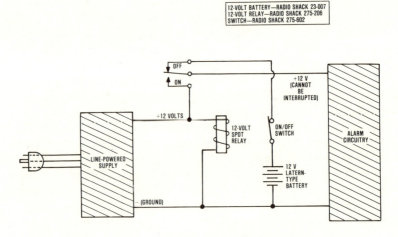

Figure 9-1. Battery back-up for a line-powered alarm

Circuit 9-1. Chemicals Cabinet Alarm

If there's a small child in your family, here's a circuit that your cleaning chemicals cabinet and medicine chest shouldn't be without. Small and compact, it fits right inside the cabinet and begins a countdown every time the door is opened. If the door is still opened at the end of an 8-12 second interval—long enough for older members of the family to conduct normal business but too

Dynamic Semiconductor Warning Mechanisms 165

short for younger members to get into trouble—it sounds a piercing tone. This noise should distract the child until you can arrive.

For construction tips, use 9-volt battery supply number 11-18 to power this circuit. Mount all components except buzzer B_1 and switch S_1 inside a small circuit box. Attach buzzer B_1 to the outside of the circuit box in an orientation for best sound broadcast, and switch S_1 to the cabinet frame so that it closes every time the door is opened.

Next, install the circuit box inside the cabinet. Use sheet metal screws or epoxy glue to fasten it securely in place. Connect S_1 into the alarm circuitry.

Test various locations around the house to insure that you can hear the alarm when it sounds. Check the battery periodically and replace it before it becomes run-down. If you must open the cabinet door for longer than the timed interval, simply depress S_1 with a finger to de-activate the mechanism.

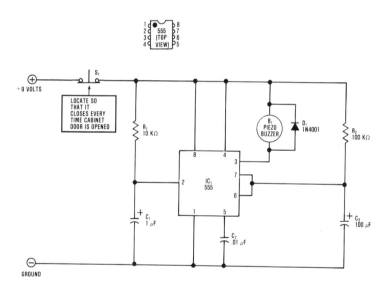

Circuit 9-1. Chemicals cabinet alarm

PARTS LIST

B_1—9-volt, low-current piezo buzzer (Radio Shack 273-060)
C_1—1-μF, 25-volt solid tantalum (Radio Shack 272-1419)
C_2—0.01-μF, 50-volt ceramic disk (Radio Shack 272-131)

C_3—100-μF, 25-volt aluminum electrolytic (Radio Shack 272-1016 or 272-1028)
D_1—1N4001 diode (Radio Shack 267-1101)
IC_1—555 integrated circuit (Radio Shack 267-1723)
R_1—10,000 ohm, ¼-watt (Radio Shack 271-1335)
R_2—100,000 ohm, ¼-watt (Radio Shack 271-1347)
S_1—Lever switch or normally closed pushbutton switch (Radio Shack 275-017 or 275-1548)

Circuit 9-2. High-Water Alarm

Flood damage brought about by an overflowing washing machine, plugged drain, broken pipe, or rain can be a thing of the past with this circuit on duty. Just locate the sensor probe in a place where seepage might first accumulate and it will "feel" for the presence of water. When this occurs, it will continuously sound a buzzer and light an LED to give you notice.

When building this circuit, first choose a buzzer of sufficient loudness for your purpose (up to 400 mA) ... then select a power supply of sufficient current capability to drive the buzzer (supply 11-4, 11-5, 11-12, or 11-20). Mount components inside a circuit box, and locate the box in a dry, central area. If you mount the buzzer

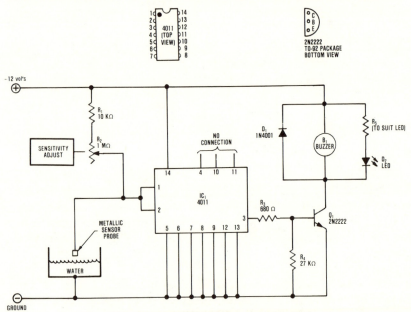

Circuit 9-2. High-water alarm

inside the circuit box, be sure to drill baffle holes in the same way as you would for a speaker. Turn "Sensitivity Adjust" R_2 to its mid-position.

Next, run a length of 2-conductor polarity-marked speaker wire between the circuit box and the area to be monitored. At the circuit box, connect the negative lead to circuit ground, and the positive lead to pins 1 and 2 of IC_1, At the area to be monitored, attach the negative lead to some low point, such as a metallic drain or a nail driven into the floor, and the positive lead to a probe suspended an inch or less above. Almost any metallic object that has sufficient mass to weight the wire (such as a nut or bolt) will do for a probe.

Finishing up, flood enough water into the monitored area to bridge the gap between the negative lead and probe. The alarm will sound as the gap is closed; become silent as the gap is opened. If necessary, rotate Sensitivity Adjust R_2 for more or less circuit sensitivity.

PARTS LIST

B_1—12-volt buzzer, up to 400 mA (Radio Shack 273-060 or 273-051)
D_1—1N4001 diode (Radio Shack 276-1101)
D_2—Any discrete red light-emitting diode (such as Radio Shack 276-026)
IC_1—4011 integrated circuit (Radio Shack 276-2411)
Q_1—2N2222 or similar transistor (Radio Shack 276-1617)
R_1—10,000 ohm, ¼-watt (Radio Shack 271-1335)
R_2—1,000,000 ohm, ¼-watt, linear taper potentiometer (Radio Shack 271-229)
R_3—680 ohm, ¼-watt (Radio Shack 271-021)
R_4—27,000 ohm, ¼-watt (Radio Shack 271-1340)
R_5—Resistor of value to suit LED

Circuit 9-3. High-Temperature Alarm

Wherever there is some maximum temperature which cannot be exceeded, this circuit can take over full-time surveillance responsibility for you. Simply set "Temperature Adjust" R_3 for the maximum allowable temperature ... position sensor Thermistor R_1 in the area to be monitored ... and this circuit will sound a tone and light an LED in the event that this preset temperature is exceeded. And it will continue to give the alarm until the problem is taken care of. Just a few of the hundreds of applications around your home or workplace include use in freezer, aquarium, greenhouse, machinery, warehouse—you can even use it as a fire alarm in conjunction with your smoke detector.

For a few tips to help you construct this circuit, first choose a buzzer of sufficient loudness for your purpose (up to 400 mA current draw)... then select a power supply capable of driving the buzzer (11-4, 11-5, 11-12, or 11-20). For use as a fire alarm, be sure to also include a battery standby as line power usually fails in a fire (Figure 9-1). Mount components inside a circuit box (except the thermistor) with front-panel accessibility for Temperature Adjust potentiometer R_3. Locate box in a protected area.

For the thermistor, almost any common 120-ohm to 150-ohm cold resistance unit will work. A unit of higher ohmic rating will also work as long as the value of resistor R_2 is increased proportionally. As a side note, "cold" resistance refers to the value at room temperature. Actual resistance of a 120-ohm unit at freezing may equal 400 ohms, and at boiling 14 ohms.

To calibrate this circuit, first rotate Temperature Adjust R_3 so that maximum voltage is seen between the wiper arm and ground (about 6 volts). Next, immerse thermistor R_1 in a container of water of the temperature at which you want the alarm to sound

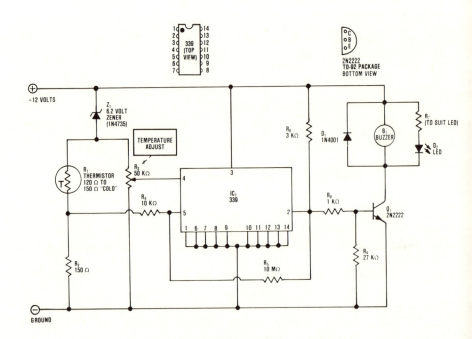

Circuit 9-3. High-temperature alarm

Dynamic Semiconductor Warning Mechanisms 169

(use a thermometer for the water). After allowing a few minutes for temperatures to settle, rotate the Temperature Adjust R_3 until the alarm just sounds. As the thermistor cools, the alarm will cease ... but as it heats to the preset temperature, the alarm will sound once again. Use 2-conductor speaker wire for interconnecting the thermistor to the monitored area. Be careful to shield the thermistor from sunlight and other sources of radiated energy which can cause the circuit to trigger prematurely.

PARTS LIST

B_1—12-volt buzzer, up to 400 mA (Radio Shack 276-060 or 273-051)
D_1—1N4001 diode (Radio Shack 276-1101)
D_2—Any discrete red light-emitting diode (such as Radio Shack 276-026)
IC_1—339 integrated circuit (Radio Shack 276-1712)
Q_1—2N2222 or similar transistor (Radio Shack 276-1617)
R_1—120- to 150-ohm "cold" resistance thermistor
R_2—150 ohm, ½-watt (Radio Shack 271-013)
R_3—50,000 ohm, ¼-watt, linear taper potentiometer (Radio Shack 271-1716)
R_4—10,000 ohm, ¼-watt (Radio Shack 271-1335)
R_5—10,000,000 ohm, ¼-watt (Radio Shack 271-1365)
R_6—3,000 ohm, ¼-watt (Radio Shack 271-1328 is OK)
R_7—Resistor of value to suit LED
R_8—1,000 ohm, ¼-watt (Radio Shack 271-1321)
R_9—27,000 ohm, ¼-watt (Radio Shack 271-1340)
Z_1—6.2-volt, 1-watt zener diode (1N4735) (Radio Shack 276-561)

Circuit 9-4. Low-Temperature Alarm

Wherever there is some minimum temperature which cannot be fallen below, this circuit can stand guard for you at the cost of only a few cents worth of energy a day. Just set "Temperature Adjust" R_3 for the minimal allowable temperature (anywhere between boiling and freezing) ... position Thermistor R_2 in the area to be monitored ... and this circuit will sound a buzzer and light an LED should the temperature fall below this limit. And it will continue to give the alarm until the problem is corrected. Just a few of the hundreds of possible applications around home, farm, and industry include use in pet cages, greenhouse, nursery, fruit orchard, and for chemical processes.

For construction, first choose a buzzer loud enough to be easily heard from where you will be (up to 400 mA) ... then select a power supply with enough current capability to drive it (supply 11-

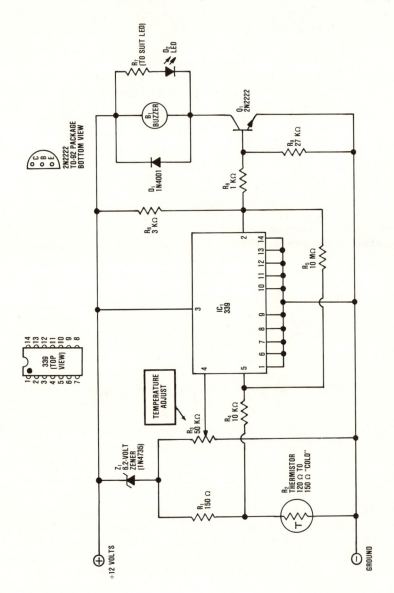

Circuit 9-4. Low-temperature alarm

Dynamic Semiconductor Warning Mechanisms 171

4, 11-5, 11-12, or 11-20). Mount components in a circuit box with front-panel accessibility for "Temperature Adjust" potentiometer R_3, and position the box in a dry protected area. If you mount the buzzer inside the circuit box, of course, cut a baffle hole as you would for a speaker.

As for the thermistor, almost any of the common 120-ohm to 150-ohm "cold" resistance units will work. A thermistor with an even higher cold resistance rating will work as long as you increase the value of resistor R_1 proportionally.

To calibrate this circuit, first turn Temperature Adjust R_3 so that maximum voltage is seen between the wiper arm and ground (about 6 volts). Next, immerse thermistor R_2 in a container of water of the temperature at which you want the alarm to sound (use a thermometer for the water). After allowing a few minutes for the temperature to settle, rotate Temperature Adjust R_3 until the alarm just sounds. As thermistor R_2 warms, the alarm will cease ... but as it cools to the preset temperature, the alarm will sound once again. Next, position this sensor in the area to be monitored, using 2-conductor speaker wire to make the interconnection. Try to position thermistor R_2 away from sunlight and other sources of radiated energy which can cause this circuit to trigger later than it normally would.

PARTS LIST

B_1—12-volt buzzer, up to 400 mA (Radio Shack 273-060 or 273-051)
D_1—1N4001 diode (Radio Shack 276-1101)
D_2—Any discrete red light-emitting diode (such as Radio Shack 276-026)
IC_1—339 integrated circuit (Radio Shack 276-1712)
Q_1—2N2222 or similar transistor (Radio Shack 276-1617)
R_1—150 ohm, ½-watt (Radio Shack 271-013)
R_2—120- to 150-ohm "cold" resistance thermistor
R_3—50,000 ohm, ¼-watt, linear taper potentiometer (Radio Shack 271-1716)
R_4—10,000 ohm, ¼-watt (Radio Shack 271-1335)
R_5—10,000,000 ohm, ¼-watt (Radio Shack 271-1365)
R_6—3,000 ohm, ¼-watt (Radio Shack 271-1328 is OK)
R_7—Resistor of value to suit LED
R_8—1,000 ohm, ¼-watt (Radio Shack 271-1321)
R_9—27,000 ohm, ¼-watt (Radio Shack 271-1340)
Z_1—6.2-volt, 1-watt zener diode (1N4735) (Radio Shack 276-561)

Circuit 9-5. Break-beam Alarm

Here's a circuit to help you keep track of the comings and goings of family members, friends, customers, pets—you can even use it for security purposes. Almost any steady light source can be used to establish an ambient level of illumination across photo cell R_2. Then when anyone steps between light and cell, this circuit will momentarily sound a small buzzer. Tone is adjustable from a friendly chirp to a shrill whine. Sensitivity may be adjusted so fine that even the slightest shadow will cause the buzzer to sound.

For construction hints, although almost any steady source of illumination may be used with this circuit, a miniature spotlight works best. You can make one with a discarded flashlight reflector and a 12-volt flashlight bulb (such as a Radio Shack 272-1143). Power supply 11-4, 11-12, or 11-20 will power both spotlight and alarm circuitry.

The best locations for photo cell R_2 and buzzer B_1 depend on your particular application. Either or both may be mounted inside the same circuit box used to house the other components via holes to admit light and emit sound. As another option, either or both

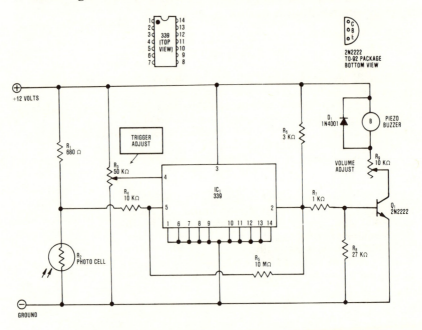

Circuit 9-5. Break-beam alarm

Dynamic Semiconductor Warning Mechanisms

may be located remotely using 2-conductor speaker wire for the interconnection with the circuitry. Note that piezo buzzers *do* have a polarity requirement. Be sure to locate "Trigger Adjust" potentiometer R_3 and "Volume Adjust" variable resistor R_9 for front-panel accessibility.

To calibrate this circuit, first rotate trigger adjust R_3 so that maximum voltage is seen between wiper and ground (12 volts). Then, after insuring that the light source and photo cell are securely fastened in their permanent positions, rotate this same control until the buzzer sounds continuously. Next, Rotate Volume Adjust R_9 for the desired volume of sound ... and then back off on Trigger Adjust R_3 until the buzzer is silent. You may want to back off on this adjustment further in order to make the alarm less sensitive.

PARTS LIST

B_1—12-volt piezo buzzer (Radio Shack 273-060)
D_1—1N4001 diode (Radio Shack 276-1101)
IC_1—339 integrated circuit (Radio Shack 276-1712)
Q_1—2N2222 or similar transistor (Radio Shack 276-1617)
R_1—680 ohm, ½-watt (Radio Shack 271-021)
R_2—Cadmium sulfide photo cell (Radio Shack 276-116)
R_3—50,000 ohm, ¼-watt linear taper potentiometer (Radio Shack 271-1716)
R_4—10,000 ohm, ¼-watt (Radio Shack 271-1335)
R_5—10,000,000 ohm, ¼-watt (Radio Shack 271-1365)
R_6—3,000 ohm, ¼-watt (Radio Shack 271-1328 is OK)
R_7—1,000 ohm, ¼-watt (Radio Shack 271-1321)
R_8—27,000 ohm, ¼-watt (Radio Shack 271-1340)
R_9—10,000 ohm, audio-taper potentiometer (Radio Shack 271-1721)

Circuit 9-6. A Complete Home Alarm System

This master-control circuit can serve as the basis for your complete home burglar-protection system. Normally closed switches S_a through S_d are a fail-safe chain of detectors such as magnetic door switches and window foils which surround your living perimeter. Should an intruder cause any of these switches to open (or cut a wire!), this circuit will activate a loud sounder in ON/OFF blasts for a full minute (for more alarm ON time, increase the value of capacitor C_3). Normally opened switches S_1 through S_4 are individually run detectors such as under-the-rug trip switches within your living perimeter. And closing any one of these switches has the same effect. Alarm ON time is adjustable from one second for testing purposes to one minute for normal service. An LED glows to tell you when this circuit has been "fired."

For construction, first select a 12-volt sounder such as a bell, blast horn, siren-horn ... then choose a power supply capable of driving it (11-5, 11-7, or 11-12 will work in most cases). Mount components in a circuit box with front-panel accessibility for "Alarm ON Time Adjust" variable resistor R_6, switch S_5, and LED D_2. Install banana jacks for plugging the detectors into the circuit (any number of detectors may be incorporated into the system). Position lock switch S_6, in a convenient outside location such as under the doorbell button. If you also want to activate the alarm while you are inside, shunt Lock Switch S_6 with a front-panel mounted ON/OFF switch. Use the relay contacts as a switch to turn ON the sounder, which is wired directly into the power supply. For extra reliability with a plug-in type supply, you may want to incorporate a battery-back-up system like the one illustrated in Figure 9-1 (on page 164).

For test-out and final adjustment, first rotate Alarm ON Time Adjust R_6 to its zero ohms position for a one-second alarm ON time. Then, with lock switch S_6 closed, trigger each detector switch in turn to insure that it fires the alarm. Note that LED D_2 glows to indicate that the alarm has been fired, and that the push-to-reload switch S_5 must be pressed or else lock switch S_6 opened and then closed before the alarm will fire again. As the last step, rotate Alarm ON Time Adjust R_6 to its full-resistance position for a one-minute alarm ON time.

PARTS LIST

C_1—4.7-μF, 25-volt aluminum electrolytic (Radio Shack 272-1012 or 272-1024)
C_2—0.1-μF, 50-volt ceramic disk (Radio Shack 272-135)
C_3—100-μF, 25-volt aluminum electrolytic (Radio Shack 272-1016 or 272-1028)
C_4—1-μF, 25-volt, solid tantalum (Radio Shack 272-1419)
C_5, C_6—0.01-μF, 50-volt ceramic disk (Radio Shack 272-131)
C_7—47-μF, 25-volt aluminum electrolytic (Radio Shack 272-1015 or 272-1027)
C_8—470-μF, 25-volt aluminum electrolytic (Radio Shack 272-1018 or 272-1030)
D_1, D_3—1N4001 diodes (Radio Shack 276-1101)
D_2—Any discrete red light-emitting diode (such as Radio Shack 276-026)
IC_1—556 integrated circuit (Radio Shack 276-1728)
Q_1—2N2222 or similar transistor (Radio Shack 276-1617)

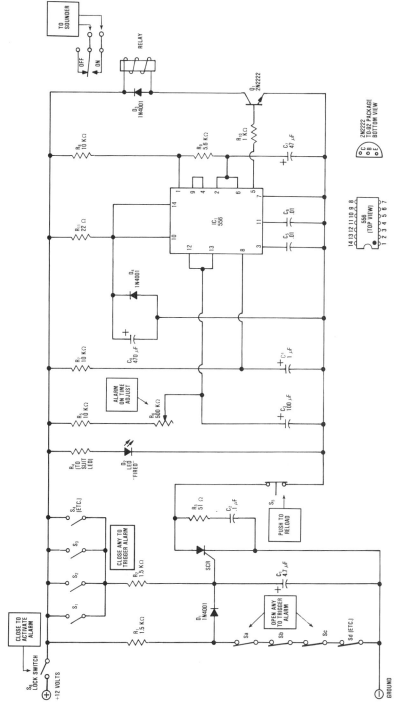

Circuit 9-6. A complete home alarm system

SCR-Any medium-current silicon-controlled rectifier (such as Radio Shack 276-1067)
Relay—12-volt relay, 250 ohms or more coil resistance, SPDT (Radio Shack 275-206)
R_1, R_2—1,500 ohm, ¼-watt (Radio Shack 271-025)
R_3—51 ohm, ¼-watt (Radio Shack 271-1307 is OK)
R_4—To suit LED
R_5, R_7, R_8—10,000 ohm, ¼-watt (Radio Shack 271-1335)
R_6—500,000 ohm, ¼-watt, linear taper potentiometer (Radio Shack 271-210)
R_9—5,600 ohm, ¼-watt (Radio Shack 271-031)
R_{10}—1,000 ohm, ¼-watt (Radio Shack 271-1321)
R_{11}—22 ohm, ½-watt (Radio Shack 271-005)
S_a through S_d—Normally closed switches (magnet switches, window foil, etc.) (check Radio Shack stock)
S_1 through S_4—Normally opened switches (panic switch, etc.) (check Radio Shack stock)
S_5—Normally closed pushbutton switch (Radio Shack 275-1548)
S_6—Lock switch (Radio Shack 49-515)

Circuit 9-7. Touch Alarm (Use In Conjunction With Timer Circuit 9-9)

Just connect this circuit to almost any metallic object and it will activate a sounder for a timed interval when this object is touched. You can have it stand watch over valued possessions such as a camera, stereo, or computer. Or, you can connect it to a doorknob and it will let you know when someone grasps the knob from the other side. You can even attach it to a file cabinet containing valuable documents...items in a display cabinet...to various objects throughout the room which an intruder would likely touch. Its uses are as limitless as your imagination!

For assembly, follow instructions for building timer circuit 9-9 and then add this circuit as the sensing mechanism. Be sure to locate "Arm Switch" S_1 and "Sensitivity Adjust" variable resistor R_1 for front-panel accessibility. If you plan to use a plug-in type power supply, you may want to include a battery back-up system (see Figure 9-1) to prevent this alarm from falsely triggering in the event that your house power fails and is then restored.

For the initial calibration of this circuit, first attach the sensor wire(s) to the object(s) to be protected. For this purpose, use alligator clips and be sure to connect to bare, clean metal. Next, open Arm Switch S_1, attach a dc voltmeter to the output (as

Dynamic Semiconductor Warning Mechanisms 177

shown), and energize the circuit. Now, rotate Sensitivity Adjust R_1 until the output voltage just goes high (about 12 volts)... then back off on this adjustment until the output voltage just goes low (about 0 volts). The output will now go high as the protected object is touched, but remain low at all other times.

Finishing up, attach the output terminal of this circuit to the input terminal of Circuit 9-9. For normal operation of the alarm, begin by placing Arm Switch S_1 in the standby position (open). Next, energize the circuit and wait at least one second for voltages to settle. Then place Arm Switch S_1 in the armed position (closed) and the alarm is in service.

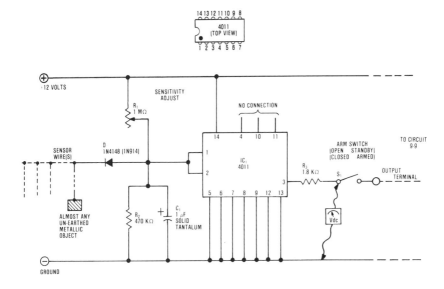

Circuit 9-7. Touch alarm (use with Circuit 9-9)

PARTS LIST

D_1—1N4148 (1N914) diode (Radio Shack 276-1122)
C_1—1-μF, 25-volt, solid tantalum (Radio Shack 272-1419 is OK)
IC_1—4011 integrated circuit (Radio Shack 276-2411)
R_1—1,000,000 ohm, linear taper potentiometer (Radio Shack 271-211)
R_2—470,000 ohm, ¼-watt (Radio Shack 271-1354)
R_3—1,800 ohm, ¼-watt (Radio Shack 271-1324)
S_1—ON/OFF switch (Radio Shack 275-602)

Circuit 9-8. Light-Detecting Alarm (Use In Conjunction With Timer Circuit 9-9)

This sophisticated alarm can protect your office desk, important filing cabinet, parts locker, tool chest—any normally shut enclosure—from being opened without your permission. Just locate photo-cell R_1 inside the enclosure to be protected, and activate the circuitry. Then when an unauthorized person opens the enclosure (thereby admitting light), this alarm will activate a sounder for a timed interval.

For assembly, follow instructions for building timer Circuit 9-9 and then add this circuit as the sensing mechanism. Be sure to locate "Sensitivity Adjust" potentiometer R_3 for front-panel accessibility. To interconnect photo cell R_1 between this circuit and the area to be monitored, use either 2-conductor speaker wire or shielded microphone cable as the case may warrant. Most photo cells have no polarity requirement. Position photo-cell R_1 inside the cabinet in a location away from holes or cracks where light seepage might otherwise cause the alarm to falsely trigger.

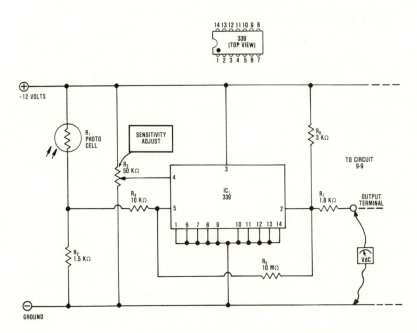

Circuit 9-8. Light-detecting alarm

Dynamic Semiconductor Warning Mechanisms 179

For the initial calibration of this circuit, leave the output terminal disconnected.

For the initial calibration of this circuit, close the monitored cabinet and attach a dc voltmeter to the output terminal (as shown). Next, energize the circuit and rotate Sensitive Adjust R_3 until the output voltage just goes high (about 12 volts) ... then back off on this adjustment until it goes—and remains—low (about 0 volts). The output should now go high when the protected cabinet is opened, but remain low at all other times.

Finishing up, attach the output terminal of this circuit to the input terminal of Circuit 9-9, and the alarm system is complete.

PARTS LIST

IC_1—339 integrated circuit (Radio Shack 276-1712)
R_1—Cadmium sulfide photo cell (Radio Shack 276-116)
R_2—1,500 ohm, ¼-watt (Radio Shack 271-025)
R_3—50,000 ohm, ¼-watt, linear taper potentiometer (Radio Shack 271-1716)
R_4—10,000 ohm, ¼-watt (Radio Shack 271-1335)
R_5—10,000,000 ohm, ¼-watt (Radio Shack 271-1365)
R_6—3,000 ohm, ¼-watt ((Radio Shack 271-1328 is OK)
R_7—1,800 ohm, ¼-watt (Radio Shack 271-1324)

Circuit 9-9. General Purpose Timer (For Use With Circuits 9-7 and 9-8)

This general-purpose timer may be used in conjunction with almost any sensing mechanism which has a normally low output voltage. Timer ON period is adjustable from one second (for testing purposes) to one minute (for normal service) (for more alarm ON time, use a larger capacitor for C_3). Power-switching ability is limited only by the current rating of the relay contacts which are used, and an LED glows to indicate when this circuit has been "fired."

For construction, first decide what type of device this timer will activate. If you plan to use the relay contacts to switch ON— for example—a tape recording of a watch dog, or a camera, then low-current power supply 11-4 or 11-12 should suffice. But if you plan to install a powerful 12-volt sounder, then a higher-current supply such as 11-5, 11-7, or 11-20 will have to be used. Mount components inside a circuit box with front-panel accessibility for

"Alarm ON Time Adjust" variable resistor R_4, "Fired" LED D_1, and "Push To Reload" switch S_1.

For testing and adjustment, first rotate Alarm ON Time Adjust R_4 to its zero-ohm position to adjust for a one-second alarm ON time. Trigger this timer via a "high" voltage (i.e., touch the input terminal to the positive rail), and check for the relay to close/the LED to glow/the alarm device to be activated. Note that once this timer has been fired, "Push To Reload" switch S_1 must be pressed or else circuit power turned OFF and then back ON again before it can be fired again. LED D_1 serves to indicate this status.

Finishing up, rotate Alarm ON Time Adjust R_4 to its maximum-ohm position to adjust for a one-minute alarm ON time (for normal service).

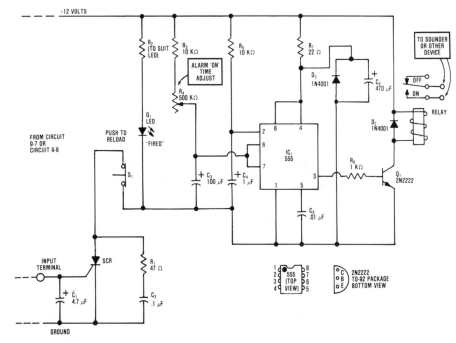

Circuit 9-9. General-purpose timer (for use with Circuits 9-7 and 9-8)

Dynamic Semiconductor Warning Mechanisms 181

PARTS LIST

C_1—4.7-μF, 25-volt aluminum electrolytic (Radio Shack 272-1012 or 272-1024)
C_2—0.1-μF, 50-volt ceramic disk (Radio Shack 272-135)
C_3—100-μF, 25-volt aluminum electrolytic (Radio Shack 272-1016 or 272-1028)
C_4—1-μF, 25-volt solid tantalum (Radio Shack 272-1419 is OK)
C_5—0.01-μF, 50-volt ceramic disk (Radio Shack 272-131)
C_6—470μF, 25-volt aluminum electrolytic (Radio Shack 272-1018 or 272-1030)
D_1—Any discrete, red light-emitting diode (such as Radio Shack 276-026)
D_2, D_3—1N4001 diode (Radio Shack 276-1101)
IC_1—555 integrated circuit (Radio Shack 276-1723)
Q_1—2N2222 or similar transistor (Radio Shack 276-1617)
Relay—12-volt relay, SPDT, 250 ohms or more coil resistance (Radio Shack 275-206)
R_1—47 ohm, ¼-watt (Radio Shack 271-1307)
R_2—To suit LED
R_3, R_5—10,000 ohm, ¼-watt (Radio Shack 271-1335)
R_4—500,000 ohm, linear taper potentiometer (Radio Shack 271-210)
R_6—1,000 ohm, ¼-watt (Radio Shack 271-1321)
R_7—22 ohm, ½-watt (Radio Shack 271-005)
SCR—Any medium-current silicon controlled rectifier (such as Radio Shack 276-1067)
S_1—Normally closed push-button switch (Radio Shack 275-1548)

In closing, why worry about the occurrence of common problems in your everyday life when you can build efficient alarm mechanisms which can actually do something about them? Each alarm shown in this chapter could possibly save you many times over the cost of its construction by giving you advance warning of a single trouble spot.

10

Using Semiconductor Circuits for Testing and Measurement Purposes at Your Workbench

There's hardly a more direct way to increase the productivity and decrease the overhead cost of your workbench than to construct your own test and measurement equipment. Of course, such complex instruments as oscilloscopes and multimeters are best bought preassembled or else in kit form. But still, you can construct a wide variety of simple yet effective electronic tools with the diagrams found in this chapter.

One example of this is the Hi/Low Logic Tester of Circuit 10-1. Simply touch the test probe to any point in the digital circuit under test, and it will "beep" if this point is high, or else remain silent if it is low. You need never take your eyes off your work! And in order to keep parts count ultra-low, this unit clips into the circuit under test for its power.

Another simplistic circuit which can be a real timesaver is Capacitance Box Circuit 10-2. For designing circuits or to replace an unmarked capacitor in an existing circuit, just clip it into the unit under test and dial-in varying amounts of capacitance until best performance is obtained. Neat and easy!

And when you are experimenting and designing, that familiar smell of overheated silicon can be a thing of the past with the Inline Current Limiter of Circuit 10-3 in place. Simply clip it between the positive lead of your power supply and the circuit under development. Then, when you accidently cause a short circuit, it will automatically limit current to a safe, preset amount (from 60 mA to about 1 amp) thereby saving delicate components from harm.

Another handy general-purpose tool is the Buzz Box shown in Circuit 10-4. In fact, you'll probably find yourself using it in lieu of your commercial multimeter in many instances. Rather than having to switch back and forth between ranges and watch a meter face, this circuit needs no adjustment or watching as it sounds an audio tone for indication. Excellent for troubleshooting shorts and opens as well as for evaluating the condition of many types of discrete components.

Other useful circuits in this chapter will produce sine waves and square waves at the frequencies which technicians find the most useful. These circuits often can eliminate the need for the more elaborate (and expensive) commercial versions.

And lastly, Capacitance Tester Circuit 10-7 can put those unmarked capacitors in your junk box back into service ... plus test capacitors for an open or a shorted condition. Just clip the capacitor to be tested into this circuit, and it will give you a reading in µF to within a very serviceable accuracy.

Circuit 10-1. Hi/Low Logic Tester

Here's an ultra-low parts circuit which can greatly speed up logic circuit troubleshooting tasks for you because you need never take your eyes off your work. It "beeps" if the test probe touches a point in the circuit under test which is high... remains silent if the point is low... and warbles if the point is oscillating between high and low. Needs no power supply of its own as it clips into the circuit under test for electrical energy. And with over 2,000,000 ohms of input resistance, it does not degrade logic levels.

For construction, mount circuit components in a miniature circuit box. Be sure to include output jacks for the power supply clip connections and for the test probe. Color-coated vinyl sheaths are recommended for the clip connections for easy polarity identification and to avoid accidental short circuits. The test probe should have a tapering and finely pointed tip in order to make easy contact with IC pins.

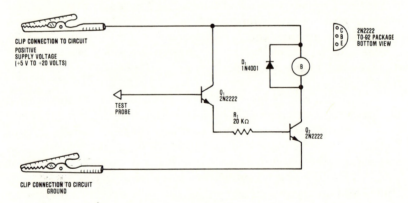

Circuit 10-1. Hi/Low logic tester

PARTS LIST

B_1—5-volt to 20-volt piezo buzzer (Radio Shack 273-060)
D_1—1N4001 diode (Radio Shack 276-1101)
Insulated Alligator Clips—(Radio Shack 270-354)
Q_1, Q_2—2N2222 or similar transistors (Radio Shack 276-1617)
R_1—20,000 ohm, ¼-watt (Radio Shack 271-1339 is OK)
Test Probe—(Radio Shack 278-740 or 278-750)

Circuit 10-2. Capacitance Box

For designing circuits or when replacing an unmarked capacitor during repair, here's a real time-saver. Simply clip it into the circuit under test, and then dial-in varying amounts of capacitance until best performance is obtained. Capacitance values include the 12 most popular sizes, ranging from 0.1-μF through 3,000-μF. The 35-volt dielectric ratings of these capacitors suit almost all semiconductor circuit requirements.

For construction, mount rotary switch S_1 and the capacitors inside a circuit box. Be sure to mark each switch position for the value of capacitance that it shunts in. Include output jacks for the test leads. Color-coded test leads are required because the electrolytic capacitors used in this unit do have a polarity requirement.

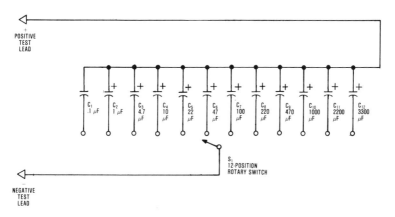

Circuit 10-2. Capacitance box

PARTS LIST

C_1—0.1-μF, 50-volt ceramic disk (Radio Shack 272-135)
C_2—1-μF, 35-volt aluminum electrolytic (Radio Shack 272-1419)
C_3—4.7-μF, 35-volt aluminum electrolytic (Radio Shack 272-1012 or 272-1024)
C_4—10-μF, 35-volt aluminum electrolytic (Radio Shack 272-1013 or 272-1025)
C_5—22-μF, 35-volt aluminum electrolytic (Radio Shack 272-1014 or 272-1026)
C_6—47-μF, 35-volt aluminum electrolytic (Radio Shack 272-1015 or 272-1027)
C_7—100-μF, 35-volt aluminum electrolytic (Radio Shack 272-1016 or 272-1028)
C_8—220-μF, 35-volt aluminum electrolytic (Radio Shack 272-1017 or 272-1029)

C_9—470-μF, 35-volt aluminum electrolytic (Radio Shack 272-1018 or 272-1030)
C_{10}—1,000-μF, 35-volt aluminum electrolytic (Radio Shack 272-1019 or 272-1032)
C_{11}—2,200-μF, 35-volt aluminum electrolytic (Radio Shack 272-1020)
C_{12}—3,300-μF, 35-volt aluminum electrolytic (Radio Shack 272-1021)
S_1—12-position rotary switch (Radio Shack 275-1385)
Test Leads—Test Lead/Clip Set (Radio Shack 270-332)

Circuit 10-3. Inline Current Limiter

Worth its weight in smoking parts! For designing and experimenting, simply connect this circuit inline with the positive output lead of your bench supply and the circuit under test, and set the rotary switch for a current limit below the rating of your weakest circuit component. Then, if an accidental short occurs (such as when a diode is installed backwards), current is held to a safe level and an LED glows to inform you that a wrong connection has been made. Operates completely independent of circuit voltage. As an added bonus, it protects against most forms of transient voltage damage.

For construction tips, mount components inside a circuit box with front-panel accessibility for rotary switch S_1. Use color-coded output jacks for connecting this unit into the circuit to be protected as the transistors have a polarity requirement. Heat-sink transistor Q_1 as it may have to dissipate considerable heat under certain conditions. Insure that all wiring and connections can easily conduct one amp of current.

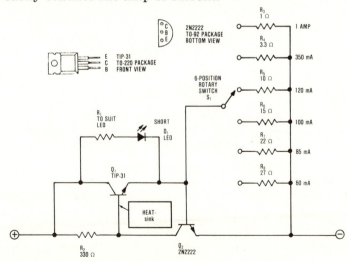

Circuit 10-3. Inline current limiter

PARTS LIST

D_1—Any red light-emitting diode (such as Radio Shack 276-026)
Q_1—TIP-31 or similar transistor (Radio Shack 276-2017)
Q_2—2N2222 or similar transistor (Radio Shack 276-1617)
R_1—To suit LED
R_2—330 ohm, ½-watt (Radio Shack 271-017)
R_3—1 ohm, 1-watt (Radio Shack 271-073)
R_4—3.3 ohm, 1-watt (Radio Shack 271-075)
R_5—10 ohm, ¼-watt (Radio Shack 271-1301)
R_6—15 ohm, ¼-watt (Radio Shack 271-003)
R_7—22 ohm, ¼-watt (Radio Shack 271-005)
R_8—27 ohm, ¼-watt (Radio Shack 271-006)
S_1—6-position rotary switch (Radio Shack 275-1385 is OK)

Circuit 10-4. Buzz Box

A simple test instrument which actually out-performs a professional-type ohmmeter in many practical test situations. You need never take your eyes off your work! Sounds a clear tone to indicate continuity, a progressively more muted tone for resistance values up to about 10,000 ohms. Checks transformers, coils, and relays quickly and accurately. A real plus for tracking-down opens and shorts in wiring. And it can be a highly effective tool for evaluating the condition of diodes, capacitors, and bi-polar transistors.

For construction, you can either use battery supply number 11-18 for portable operation of this circuit (recommended), or else line-powered supply 11-4, 11-11, or 11-12 for use as a bench instrument. Mount "Volume Adjust" variable resistor R_3 so that it is easily accessible from the front of the circuit box. Be sure to use color-coded test probes as polarity is important in many test situations.

When using this circuit for making continuity/approximate resistance measurements, loudness of tone indicates approximate value of resistance shunted across the test probes (tone ceases at about 10,000 ohms). For locating shorts/opens in wiring, simply clip the test probes across the faulty circuit and separate wires/tug at connections while listening for tone to cease/start. Sense and operation of diodes and bi-polar transistors are quickly tested by touching the *n*egative probe to N-type material, the *p*ositive probe to P-type material, and listening for the resultant tone as electrical current flows through the P-N junction (no tone occurs if the probes are reversed).

To test capacitors (out of circuit) for normal operation (1-μF or greater), first short the capacitor's leads together in order to equalize the charge across its plates. Then shunt the test probes across the capacitor (observing polarity if required) while listening to the loudness and duration of the resultant tone. Check several capacitors that you know to be good/bad for comparison purposes.

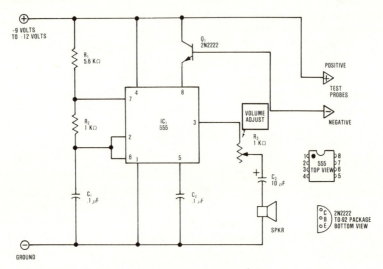

Circuit 10-4. Buzz box

PARTS LIST

C_1, C_2—.1-μF, 50-volt ceramic disk (Radio Shack 272-135)
C_3—10-μF, 25-volt aluminum electrolytic (Radio Shack 272-1013 or 272-1025)
IC_1—555 integrated circuit (Radio Shack 276-1723)
Q_1—2N2222 or similar transistor (Radio Shack 276-1617)
R_1—5,600 ohm, ¼-watt (Radio Shack 271-031)
R_2—1,000 ohm, ¼-watt (Radio Shack 271-1321)
R_3—1,000 ohm, ¼-watt (or more), linear taper potentiometer
SPKR—Any 8-ohm speaker
Test Probes—(Radio Shack 278-740 or 278-750)

Circuit 10-5. Variable-Frequency Square-Wave Generator

A low-parts yet highly effective means of generating square waves for use as clock pulses or for test and evaluation purposes. Output frequency is continually variable from about 40 Hz. through 40K Hz. Waveforms are very symmetrical up to about 10

Circuits for Testing and Measurements

KHz. Wide voltage supply range allows use with most semiconductor circuits. Drives any circuit with 2,000 or more ohms of input resistance.

For tips to help you construct this circuit, use battery supply 11-17 or 11-18 for portable operation, or line-powered supply 11-1, 11-4, 11-6, 11-10, 11-11, 11-12, 11-13, or 11-14 to avoid periodic battery replacement. As another option, you may power this circuit from your bench-type power supply. Mount "Frequency Adjust" variable resistor R_2 and "Range Adjust" switch S_1 for easy access from the front of the circuit box.

One useful purpose of this circuit is to test audio amplifiers for frequency response. To do this, inject a square wave into the input of the amplifier while you monitor the output of the amplifier with an oscilloscope (be sure to use a dummy load for the

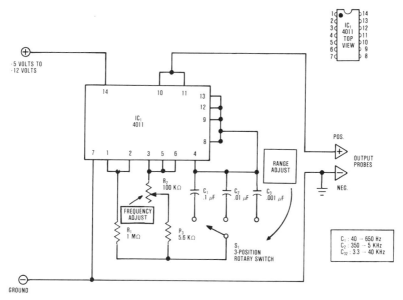

Circuit 10-5. Variable-frequency square-wave generator

PARTS LIST

C_1—0.1-μF, 50-volt ceramic disk (Radio Shack 272-135)
C_2—0.01-μF, 50-volt ceramic disk (Radio Shack 272-131)
C_3—0.001-μF, 50-volt ceramic disk (Radio Shack 272-126)
IC_1—4011 integrated circuit (Radio Shack 276-2411)
R_1—1,000,000 ohm, ¼-watt (Radio Shack 271-1356)
R_2—100,000 ohm, ¼-watt (Radio Shack 271-1347)
R_3—5,600 ohm, ¼-watt (Radio Shack 271-031)
S_1—3-position rotary switch (Radio Shack 275-1385 is OK)

amplifier output in order to avoid a possible burnout). The frequency at which the square-wave pattern has notably rounded edges relates to about 1/3 the cutoff frequency of the amplifier. For example, if square-wave frequencies of 6 KHz. and above have rounded transition edges as seen on the oscilloscope, then the cutoff frequency of the amplifier is about 18,000 Hz.

Circuit 10-6. Audio-Signal Generator

Produces a 3K Hz. sine-wave signal for such purposes as gain measurement in audio equipment. Drives any load with 10,000 or more ohms of resistance, although a direct short on the output of this circuit does no harm. And with so few parts, it's extraordinarily easy to construct.

For constructing this circuit, use battery supply 11-18 for portable operation, or plug-in supply 11-4, 11-11, or 11-12 to avoid battery replacement. As a parts-saving option, you can power this circuit with your bench supply. Be careful to keep component lead lengths short in order to thwart harmonic oscillations. For increased testing versatility, this circuit may be mounted in the same circuit box and powered by the same supply as square-wave generator Circuit 10-5.

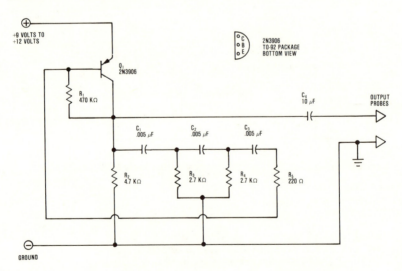

Circuit 10-6. Audio-signal generator

Circuits for Testing and Measurements 193

PARTS LIST

C_1, C_2, C_3—.005-μF, 50-volt ceramic disk (Radio Shack 276-130)
C_4—10-μF, 25-volt, non-polarized aluminum electrolytic (Radio Shack 272-999)
Q_1—2N3906 or similar transistor (Radio Shack 276-2034)
R_1—470,000 ohm, ¼-watt (Radio Shack 271-1354)
R_2—4,700 ohm, ¼-watt (Radio Shack 271-1330)
R_3, R_4—2,700 ohm, ¼-watt (Radio Shack 271-1328 is OK)
R_5—220 ohm, ¼-watt (Radio Shack 271-1313)

Circuit 10-7. Capacitance Tester

Identifying the value of unknown capacitors does not require an expensive test instrument. Here's a circuit which gives a serviceable plus or minus 20% accuracy, and can be made even more accurate if necessary. Simply clip the capacitor to be tested between Point 1 and Point 2 ... set "Series Resistance Select" switch S_1 to an estimated value ... push "Press To Test" switch S_2 ... and measure elapsed time until the buzzer sounds, using an ordinary watch. Then from series resistance and elapsed time you can easily calculate the value of capacitance using the formula:

$$\text{Capacitance } (\mu F) = \frac{\text{Time in Seconds}}{\text{Series Resistance}}$$

For construction, you may use battery supply 11-17 or 11-18 for portable operation, or line-powered supply 11-1, 11-4, 11-10, 11-11, or 11-12 for bench use. Mount Series Resistance Select switch S_1, normally closed Press-to-Test switch S_2, and Points 1 and 2 for access from the front of the circuit box. Use color-coded alligator clips for connecting capacitors between Points 1 and 2 (most electrolytic capacitors have a polarity requirement).

For even higher accuracy, substitute a 4,700-ohm resistor and a 10,000-ohm variable resistor for Resistor R_8. Then, using an accurate capacitor of known value (solid tantalum capacitors are usually quite close to rated value), calibrate this 10,000-ohm variable resistor for the correct elapsed time. Note that aluminum electrolytic capacitors commonly vary from rated value by a factor of from minus 20% to plus 50%. Use a stop watch for a more accurate measurement of time. Also note that if a capacitor fails to cause this circuit to sound after a reasonable time (or if this circuit sounds instantly regardless of series resistance), the capacitor is faulty.

194 Circuits for Testing and Measurements

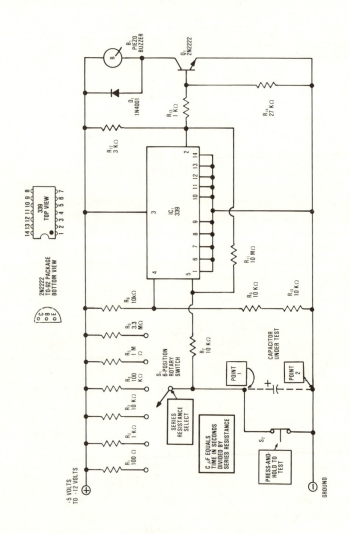

Circuit 10-7. Capacitance tester

Circuits for Testing and Measurements

PARTS LIST

B_1—5-volt to 12-volt piezo buzzer (Radio Shack 273-060)
D_1—1N4001 diode (Radio Shack 276-1101)
IC_1—339 integrated circuit (Radio Shack 276-1712)
Q_1—2N2222 or similar transistor (Radio Shack 276-1617)
R_1—100 ohms, ¼-watt, 5% tolerance (Radio Shack 271-1311)
R_2, R_{13}—1,000 ohm, ¼-watt, 5% tolerance (Radio Shack 271-1321)
R_3, R_7, R_8, R_9, R_{10}—10,000 ohm, ¼-watt, 5% tolerance (Radio Shack 271-1335)
R_4—100,000 ohm, ¼-watt, 5% tolerance (Radio Shack 271-1347)
R_5—1,000,000 ohm, ¼-watt, 5% tolerance (Radio Shack 271-1356)
R_6—3,300,000 ohm, ¼-watt, 5% tolerance
R_{11}—10,000,000 ohm, ¼-watt (Radio Shack 271-1365)
R_{12}—3,000 ohm, ¼-watt (Radio Shack 271-1328 is OK)
R_{14}—27,000 ohm, ¼-watt (Radio Shack 271-1340)
S_1—6-position rotary switch (Radio Shack 275-1385 is OK)
S_2—Normally closed push-button switch (Radio Shack 275-1548)

In closing, consider combining a number of the test circuits detailed in this chapter into a single circuit box. This will not only reduce the space they consume at your workbench, but will lower construction costs as they can all share a single power supply. But most important, this will centralize a number of valuable test instruments within quick and easy reach at your workbench.

Semiconductor Power Supplies and Auxiliary Circuits

This chapter contains detailed instructions for building the power supplies and auxiliary circuits that you will need for completion of the circuits and projects found in Chapters 2 through 10. In fact, it will show you how to build practically any supply you will ever need for any purpose. But before we begin, let's review data for selecting the best supply for your application.

Three types of power supply are listed in the following pages: integrated circuit-regulated supplies, zener diode-regulated supplies, and battery-powered supplies. Several types are recommended for use with most of the preceding circuits and projects. So which type should you use?

A supply built around a high-performance integrated circuit is often your most practical choice. This type of supply features a ripple rejection figure of better than 50 dB plus excellent load and line regulation. What's more, this type of supply will automatically shut itself down in the event of a short circuit across its output or of a thermal overload, and usually requires only a minimum number of circuit components. And minimum components mean reduced assembly time as well as less perfboard space consumption. As a slight drawback, IC-based supplies require the use of somewhat more expensive and less available tantalum capacitors for best performance.

A zener diode-regulated supply, on the other hand, offers some real advantages in cost and parts availability. In most cases, you can construct this type of supply entirely from the lowest-priced and most widely available parts on your dealer's shelf. Or, if you prefer, you can put one together for next to nothing using similar parts left over from previous projects or salvaged from a scrap chassis. Besides, working with discrete components rather than integrated circuits is often a refreshing change. On the minus side, however, zener supplies usually give lesser (though still very adequate) performance than comparable IC supplies.

But don't overlook battery-powered supplies. They are rugged, compact (usually), and completely do away with the hazards of working with line voltage. Actually, they are the only practical supply for use with most portable circuits. Their major disadvantage is, of course, that batteries need periodic recharging and/or replacement.

And once again before you begin, remember to be cautious when working with line voltage. The first line of defense here is plain common sense. Also, remember that most mishaps around the workbench occur toward the end of the day when technicians are tired (therefore less alert). Now is a good time to review "Playing It Safe With High Voltage," in Chapter 1.

INTEGRATED CIRCUIT-REGULATED SUPPLIES

Circuit 11-1. 5-Volt, Low-Current IC-Regulated Supply

For powering low-current TTL circuits, plus many other types of transistor circuits, this supply can be nearly ideal. Because it requires only 8 parts and 17 soldered connections, you can snap it together quickly and easily. And it consumes only a minimal amount of circuit board space.

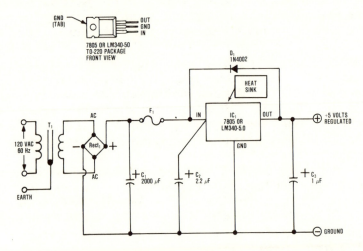

Circuit 11-1. 5-volt, low-current, IC-regulated supply

Semiconductor Power Supplies and Auxiliary Circuits 201

Under test with a 12.6-volt, 300-mA transformer and a TO-220 housed 7805 regulator, this circuit could easily supply 150 mA of current into a 33-ohm load (0.74 watts). The output voltage dropped 0.1 volts as the load was attached, which equates to a 2% regulation figure. Efficiency at full load was a little better than 16%, and no-load power draw was 3.8 watts.

For construction tips, a TO-220 housed regulator with moderate heat sinking is quite adequate for this supply. If you plan to use this supply for powering TTL circuitry, make sure that the unloaded output voltage does not exceed 5.25 volts through a tolorance error (maximum TTL voltage), nor fall below 4.75 volts under load (minimum TTL voltage).

PARTS LIST

C_1—2,000-μF, 25-volt aluminum electrolytic (Radio Shack 272-1020)
C_2—2.2-μF, 25-volt solid tantalum
C_3—1-μF, 25-volt solid tantalum
D_1—IN4002 diode (Radio Shack 276-1102)
F_1—Choose to match transformer rating
IC_1—7805 or LM340-5.0, 3-terminal voltage regulator (Radio Shack 276-1770)
$Rect_1$—1-amp, 50-PIV, modular bridge rectifier (Radio Shack 276-1161)
T_1—12-volt or 12.6-volt, up to 500-mA secondary, miniature transformer (Radio Shack 273-1385)

Circuit 11-2. 5-Volt, Medium-Current, IC-Regulated Supply

Here's a supply for use with TTL, plus many many other types of transistor circuits, that provides extra current for driving relays, lights, motors, and buzzers. And it needs no fuse, for in the event of a short circuit at its output, the IC regulator will limit current to a safe level at or below the maximum ratings of circuit components.

As tested with a 12.6-volt, 1.2 amp transformer coupled to a 7805 regulator which had a TO-220 case, this supply could easily sustain 400 mA into a 12-ohm load (1.9 watts). Voltage drop from no load to full load was 0.2 volts, which calculates to a 4% regulation figure. Efficiency under full load was slightly better than 15%, and no-load power draw from the line was 6.5 watts.

As a few helpful construction pointers, for highest current output, be sure to heat-sink the regulator well. If possible, try to use a regulator packaged in a TO-3 case with this supply. All things equal, the LM340-XX series regulator gives somewhat superior performance as compared to the 78XX series. If you plan to use this supply for powering TTL circuitry, make sure that the unloaded output voltage does not exceed 5.25 volts through a tolerance error (maximum TTL voltage), or fall below 4.75 volts under load (minimum TTL voltage).

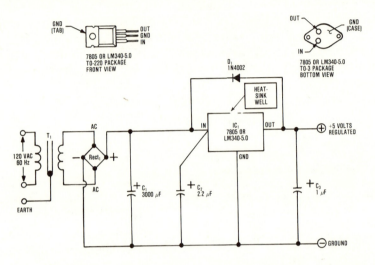

Circuit 11-2. 5-volt, medium-current, IC-regulated supply

PARTS LIST

C_1—3,000-μF, 25-volt aluminum electrolytic (Radio Shack 272-1021)
C_2—2.2-μF, 25-volt solid tantalum
C_3—1-μF, 25-volt solid tantalum
D_1—1N4002 diode (Radio Shack 276-1102)
IC_1—7805 or LM340-5.0, 3-terminal voltage regulator (Radio Shack 276-1770)
$Rect_1$—2-amp, 50-PIV, modular bridge rectifier (Radio Shack 276-1146)
T_1—12-volt or 12.6-volt, 1-amp or 1.2-amp secondary, standard transformer (Radio Shack 273-1505)

Circuit 11-3. 5-Volt, High-Current, IC-Regulated Supply

This high-performance supply provides plenty of current for larger 5-volt projects. It can output better than 1.5 amps on a

Semiconductor Power Supplies and Auxiliary Circuits 203

continuous basis, and with its high energy-conversion factor you can operate it for extended periods at very little expense on your monthly power bill.

As tested with an 18-volt, center-tapped, 2-amp, heavy-duty transformer and a 7805 regulator (TO-220 package), this supply generated an easy 1.5 amps into a 3.25-ohm load (7.4-watts). Voltage drop as the load was attached was a low 0.1 volts which

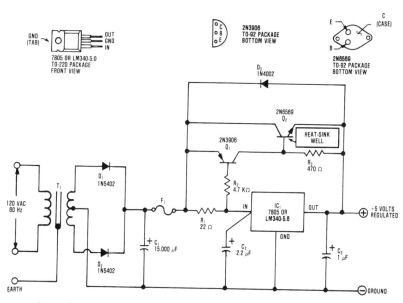

Circuit 11-3. 5-volt, high-current, IC-regulated supply

PARTS LIST

C_1—15,000-μF, 25-volt aluminum electrolytic
C_2—2.2-μF, 25-volt solid tantalum
C_3—1-μF, 25-volt solid tantalum
D_1, D_2—IN5402 diodes (Radio Shack 276-1143)
D_3—IN4002 diode (Radio Shack 276-1102)
F_1—2-amp fuse
IC_1—7805 or LM340-5.0, 3-terminal voltage regulator (Radio Shack 276-1770)
Q_1—2N3906 or similar transistor (Radio Shack 276-2034)
Q_2—2N6569 or similar transistor (Radio Shack 276-2039)
R_1—22 ohm, ¼-watt (Radio Shack 271-005)
R_2—4,700 ohm, ¼-watt (Radio Shack 271-1330)
R_3—470 ohm, ¼-watt (Radio Shack 271-1317)
T_1—18-volt, center-tapped, 2-amp, heavy-duty transformer (Radio Shack 273-1515)

calculates to a 2% regulation, and full-load efficiency was a high 31.25%. No-load power consumption was 7 watts.

As construction tips, be sure to heat-sink pass transistor Q_2 well as it heats under conditions of high-current demand. An external, finned sink is recommended. The IC regulator, however, heats very little and requires only minimal sinking. When working with high-value capacitor C_1, be sure that it is completely discharged before you handle it to avoid a spark hazard.

Circuit 11-4. 9-Volt to 12-Volt Adjustable, Low-Current, IC-Regulated Supply

When project specifications call for a 9-volt or a 12-volt low-current supply, here's an adaptable unit that can do the job handily. You can even adjust it down as low as 5 volts. Or use it for one project at a given voltage, then—with the turn of the adjustment dial—use it for another at a different voltage. And if you want to use this supply for 12-volt-only operation, you may omit several parts.

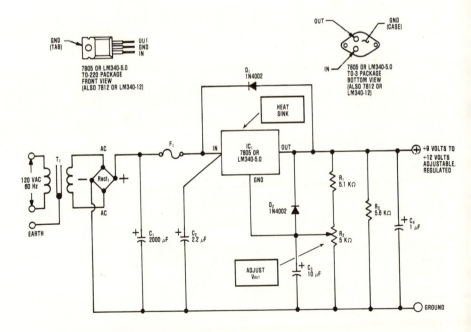

Circuit 11-4. 9-volt to 12-volt adjustable, low-current, IC-regulated supply

Under test with a 12.6-volt, 300-mA transformer, plus a 7805 regulator (TO-220 package) set for 9 volts, this supply delivered an easy 150 mA into a 58-ohm load (1.3 watts). Voltage falloff as the load was attached was 0.3 volts for a 3.4% regulation figure. Efficiency was an adequate 22.6% under full load, and no-load power draw from the line was 3.8 watts.

When set for 12 volts, this supply delivered an easy 130 mA into a 90-ohm load (1.5 watts). No-load to full-load regulation was 0.5 volts for a 4.2% regulation, and full-load efficiency was a quite good 28.5%. Ambient power draw was the same at 3.8 watts.

For construction pointers, to adjust the output voltage for your exact needs, rotate "Adjust V_{out}" potentiometer R_2. However, it you need a 12-volt-only output (non-adjustable), you may omit parts D_2, C_3, R_1, R_2, and R_3. Then substitute a 7812 or an LM340-12 regulator for the 7805/LM340-5.0, and connect the Gnd. terminal of this regulator to the negative rail of the supply.

PARTS LIST

C_1—2,000-μF, 25-volt aluminum electrolytic (Radio Shack 272-1020)
C_2—2.2-μF, 25-volt solid tantalum
C_3—10-μF, 25-volt solid tantalum
C_4—1-μF, 25-volt solid tantalum
D_1, D_2—IN4002 diodes (Radio Shack 276-1102)
F_1—Choose to match transformer rating
IC_1—7805 or LM 340-5.0, 3-terminal voltage regulator (or a 7812/LM340-12) (Radio Shack 276-1770 or 276-1771 respectively)
R_1—5,100 ohm, ¼-watt (Radio Shack 271-031 is OK)
R_2—5,000 ohm, ¼-watt, linear taper potentiometer (Radio Shack 271-217)
R_3—5,600 ohm, 1/2-watt (Radio Shack 271-031)
$Rect_1$—1-amp, 50-PIV, modular bridge rectifier (Radio Shack 276-1161)
T_1—12-volt or 12.6-volt, up to 500-mA secondary, miniature transformer (Radio Shack 273-1385)

Circuit 11-5. 9-Volt to 12-Volt Adjustable, Medium-Current, IC-Regulated Supply

For powering 9-volt to 12-volt projects which need extra current for driving lights, relays, motors, buzzers, or for other reasons, this versatile supply can fill the requirements. You can even adjust it down as low as 5 volts provided that the IC regulator has a proper heat sink. Use it for one project at a given voltage then, with a simple adjustment, use it for another project at a different voltage. As an additional advantage, if you wish to use this supply for 12-volt-only operation, you may omit a number of parts.

When equipped with a 12.6-volt, 1.2-amp transformer coupled to a TO-220 packaged IC regulator and set for 9 volts, this supply could easily output 250 mA into a 34-ohm load (2.2 watts). Voltage drop as the load was attached was 0.3 volts which calculates to a 3.4% regulation figure. Full-load efficiency was a respectable 22%, and no-load power draw from the line was 6.2 watts.

Set for 12 volts, this supply could easily deliver 280 ma into a 40-ohm load (3.1 watts). Voltage drop as the load was attached was 0.5 volts, yielding a 4.2% regulation figure. Full-load efficiency was a quite good 32%, and no-load power consumption was the same at 6.2 watts.

For pointers, using a regulator encased in a TO-3 package will give you even better performance than specified—especially at the 9-volt setting. Rotate "Adjust V_{out}" potentiometer R_2 to vary the output voltage. If you need a 12-volt-only supply (non-adjustable), you can reduce expenses by omitting components D_2, C_3, R_1, R_2, and R_3. Then substitute a 7812 or a LM340-12 regulator for the 7805/LM340-5.0 one which is specified, and connect the Gnd. terminal of this regulator to the negative rail of the supply.

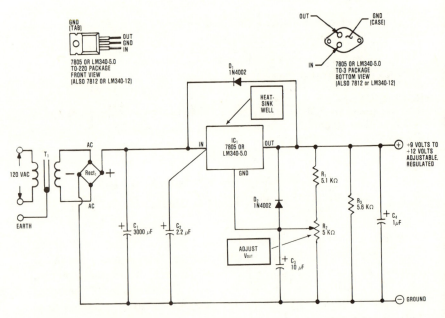

Circuit 11-5. 9-volt to 12-volt adjustable, medium-current, IC-regulated supply

Semiconductor Power Supplies and Auxiliary Circuits **207**

PARTS LIST

C_1—3,000 μF, 25-volt aluminum electrolytic (Radio Shack 272-1021)
C_2—2.2-μF, 25-volt, solid tantalum
C_3—10-μF, 25-volt solid tantalum
C_4—1-μF, 25-volt solid tantalum
D_1, D_2—1N4002 diodes (Radio Shack 276-1102)
IC_1—7805 or LM340-5.0, 3-terminal voltage regulator (or a 7812/LM340-12) (Radio Shack 276-1770 or 276-1771 respectively)
R_1—5,100 ohm, ¼-watt (Radio Shack 271-031 is OK)
R_2—5,000 ohm, ¼-watt, linear taper potentiometer (Radio Shack 271-217)
R_3—5,600 ohm, 1/2-watt (Radio Shack 271-031)
$Rect_1$—2-amp, 50-PIV, modular bridge rectifier (Radio Shack 276-1146)
T_1—12 volt or 12.6 volt, 1-amp or 1.2-amp secondary, standard transformer (Radio Shack 273-1505)

Circuit 11-6. 15-Volt, Low-Current, IC-Regulated Supply

Here's an 8-part solution to the task of providing many semiconductor circuits with a 15-volt, regulated supply. With a parts count this low, you can assemble it quickly and easily. Another plus is that it requires only a minimal amount of perfboard space for construction.

As tested with a 25.2-volt, 300-mA transformer and a TO-220-housed 7815 regulator, this supply produced an easy 150 mA

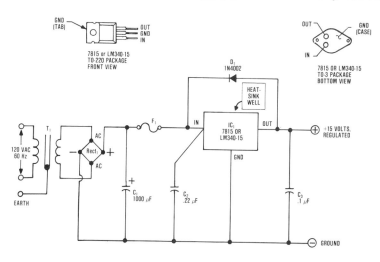

Circuit 11-6. 15-volt, low-current, IC-regulated supply

into a 100-ohm load (2.25 watts). The voltage dropped 0.2 volts from an unloaded to a fully loaded condition for a low 1.4% regulation figure, and efficiency was an adequate 19.5%. No-load ambient power consumption was measured at 8.2-watts.

For construction hints, you can get increased power efficiency at little expense to regulation by using a 20-volt or 24-volt transformer (rather than a 25.2-volt unit). If your project will draw a comparatively heavy current from this supply on a continuous basis, a TO-3 packaged regulator with a good heat sink will provide best long-term reliability.

PARTS LIST

C_1—1,000-μF, 50-volt aluminum electrolytic (Radio Shack 272-1047)
C_2—0.22-μF, 50-volt ceramic disk
C_3—0.1-μF, 50-volt ceramic disk (Radio Shack 272-135)
D_1—1N4002 diode (Radio Shack 276-1102)
F_1—Choose to match transformer rating
IC_1—7815 or LM340-15 3-terminal voltage regulator (Radio Shack 276-1772)
$Rect_1$—1-amp, 100-PIV modular bridge rectifier (Radio Shack 276-1152)
T_1—20-volt to 25.2-volt, up to 500-mA secondary, miniature transformer (Radio Shack 273-1386)

Circuit 11-7. 12-Volt to 15-Volt Adjustable, High-Current, IC-Regulated Supply

This high-performance supply is one of the most versatile and useful circuits in this chapter. It can be used to power larger projects requiring a 12-volt to 15-volt supply. Also, it's a natural for at-home use of most car radios, 8-tracks, cassette players, and CB's. You can even use it as a bench supply for servicing mobile equipment. Efficiency is excellent, too, so you can operate it for extended periods at little cost on your monthly electric bill.

As tested with an 18-volt, 2-amp transformer and a TO-220 packaged 7812 regulator which was set for 12 volts, this supply easily conducted 1.25 amps into a 9.5-ohm load (14.85-watts). The voltage drop from unloaded to fully loaded was a mere 0.1 volts yielding a regulation factor of 0.84%. Efficiency was an excellent 48% under full load, and no-load power draw came to 7.2 watts.

For construction tips, pass transistor Q_2 tends to heat under load and therefore needs good heat-sinking. A finned, external sink is recommended. The IC regulator, however, heats very little and requires minimal, if any, sinking. To adjust the output

Semiconductor Power Supplies and Auxiliary Circuits 209

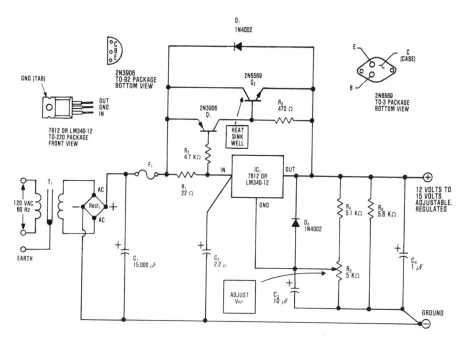

Circuit 11-7. 12-volt to 15-volt adjustable, high-current, IC-regulated supply

PARTS LIST

C_1—1,5000-μF, 35-volt aluminum electrolytic
C_2—2.2-μF, 35-volt, solid tantalum
C_3—10-μF, 25-volt solid tantalum
C_4—1-μF, 35-volt solid tantalum
D_1, D_2—1N4002 diodes (Radio Shack 276-1102)
F_1—2-amp fuse
IC_1—7812 or LM340-12 3-terminal voltage regulator (Radio Shack 276-1771)
Q_1—2N3906 or similar transistor (Radio Shack 276-2034)
Q_2—2N6569 or similar transistor (Radio Shack 276-2039)
R_1—22 ohm, ¼-watt (Radio Shack 271-005)
R_2—4,700 ohm, ¼-watt (Radio Shack 271-1330)
R_3—470 ohm, ¼-watt (Radio Shack 271-1317)
R_4—5,100 ohm, ¼-watt (Radio Shack 271-031 is OK)
R_5—5,000 ohm, ¼-watt, linear taper potentiometer (Radio Shack 271-217)
R_6—5,600 ohm, 1/2-watt (Radio Shack 271-031)
$Rect_1$—4-amp, 100-PIV, modular bridge rectifier (Radio Shack 276-1171)
T_1—18 volt, 2-amp secondary, heavy-duty transformer (Radio Shack 273-1515)

voltage, simply rotate "Adjust V_{out}" potentiometer R_5 for the desired setting. Most mobile equipment operates best at 13.8 volts, which is about the voltage is a vehicle's electrical system when the alternator is running. Remember to discharge filter capacitor C_1 thoroughly before handling it as it can store enough energy to give a potentially hazardous spark.

Circuit 11-8.
±9 Volt to ±12 Volt, Dual Complementary, IC-Regulated Supply

For circuits containing op-amps or other components requiring a dual supply, here's a smooth-operating unit that's hard to beat. Continuously adjustable plus and minus outputs allow you to trim supply voltages to exact specifications for best performance. And the use of easy-to-find stock transformers eliminates the problem of locating an expensive center-tapped unit.

As tested with dual 12.6-volt, 300-mA miniature transformers and a 7805/7905 regulator combination (TO-220 housed) which were set for 9 volts, this supply delivered an easy 120 mA into a 150-ohm load shunted across the plus/minus output terminals (2 watts). The voltage dropped 0.4 volts as the load was attached, which related to a 2.2% regulation figure. Full-load efficiency was better than 24%, and no-load power consumption was 7.6 watts.

Here are a couple of tips to help you assemble this circuit. Because this supply consists of symmetrical upper and lower halves which operate in tandem, corresponding parts should match for best performance. We've given these corresponding upper and lower parts the same number (except for an "A" suffix following the lower part number) to help you identify them. Be sure to use matched transformers from the same manufacturer, and to line them up end-to-end so that you can't make a mistake in their interconnection. For regulators, units housed in the TO-220 package are satisfactory if equipped with good heat sinks. Rotate "Adjust V_{out}" potentiometers R_3/R_{3A} to set the plus/minus output voltages (in respect to common).

Semiconductor Power Supplies and Auxiliary Circuits

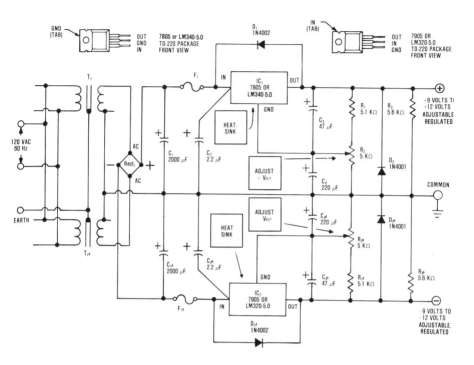

Circuit 11-8. ±9 volts to ±12 volt dual complementary, IC-regulated supply

PARTS LIST

C_1, C_{1A}—2,000-μF, 25-volt aluminum electrolytic (Radio Shack 272-1020)
C_2, C_{2A}—2.2-μF, 25-volt solid tantalum
C_3, C_{3A}—47-μF, 25-volt aluminum electrolytic (Radio Shack 272-1015 or 272-1027)
C_4, C_{4A}—220-μF, 25-volt aluminum electrolytic (Radio Shack 272-1017 or 272-1029)
D_1, D_{1A}—1N4002 diodes (Radio Shack 276-1102)
D_2, D_{2A}—1N4001 diodes (Radio Shack 276-1101)
F_1, F_{1A}—Choose to match transformer ratings
IC_1—7805 or LM340-5.0 3-terminal voltage regulator (Radio Shack 276-1770)
IC_2—7905 or LM 320-5.0 3-terminal voltage regulator (Radio Shack 276-1773)
R_1, R_{1A}—5,100 ohm, ¼-watt (Radio Shack 271-031 is OK)
R_2, R_{2A}—5,000 ohm, ¼-watt, linear taper potentiometer (Radio Shack 271-217)
R_3, R_{3A}—5,600 ohm, 1/2-watt (Radio Shack 271-031)
$Rect_1$—2-amp, 100-PIV, modular bridge rectifier (Radio Shack 276-1171)
T_1, T_{1A}—12-volt or 12.6-volt, up to 500-mA secondary, miniature transformers (Radio Shack 273-1385)

Circuit 11-9. ±15 Volts to ±20 Volts, Dual Complementary, IC-Regulated Supply

When optimum performance in op-amp and other circuitry calls for a ±15-volt to a ±20-volt supply, this one can be nearly ideal. You can trim each output voltage to exact specifications for top performance, and the use of dual transformers eliminates the need for an expensive center-tapped unit while even improving performance.

Under test with twin 25.2-volt, 300-mA transformers and TO-220 packaged 7815/7915 regulators set for 15 volts, this supply easily generated a continuous 75 mA into a 400-ohm load shunted across the plus/minus output terminals (2.25 watts). The voltage drop as the load was attached was a low 0.2 volts, which calculates to a regulation figure of 0.68%. Full-load efficiency was a little better than 10%, and no-load power consumption was measured at 15.5 watts.

For construction advice, 20-volt or 24-volt transformers are the better choice for this supply as they will give improved efficiency at little expense to regulation. Because this supply consists of symmetrical upper and lower halves which operate in tandem, each corresponding part should match for best performance. We've given these corresponding parts the same number (except for an "A" suffix following the lower part number) to help you identify them. Be sure to use matched transformers from the same manufacturer, and line them up end-to-end so that you cannot make a mistake in their interconnection. Heat-sink the regulators well. To set the plus/minus output voltages with respect to common, rotate "Adjust V_{out}" potentiometers R_2/R_{2A}.

PARTS LIST

C_1, C_{1A}—1,000-μF, 50-volt aluminum electrolytic (Radio Shack 272-1047)
C_2, C_{2A}—0.22-μF, 50-volt ceramic disk
C_3, C_{3A}—47-μF, 35-volt aluminum electrolytic (Radio Shack 272-1015 or 272-1027)
C_4, C_{4A}—220-μF, 35-volt aluminum electrolytic (Radio Shack 272-1017 or 272-1029)
D_1, D_{1A}—1N4002 diodes (Radio Shack 276-1102)
D_2, D_{2A}—1N4001 diodes (Radio Shack 272-1101)
F_1, F_{1A}—Choose to match transformer rating
IC_1,—7815 or LM340-15 3-terminal voltage regulator (Radio Shack 276-1772)
IC_2—7915 or LM320-15 3-terminal voltage regulator
R_1, R_{1A}—5,100 ohm, ¼-watt (Radio Shack 271-031 is OK)

Semiconductor Power Supplies and Auxiliary Circuits

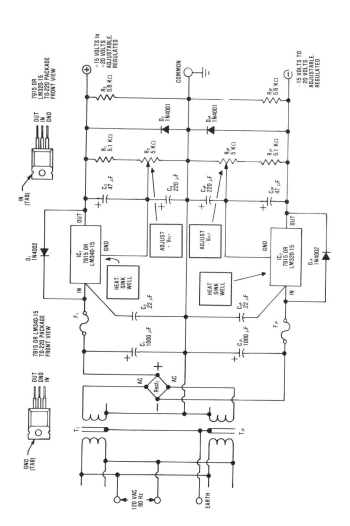

Circuit 11-9. ±15 volts to ±20 volt dual complementary, IC-regulated supply

214 Semiconductor Power Supplies and Auxiliary Circuits

R_2, R_{2A}—5,000 ohm, ¼-watt, linear taper potentiometer (Radio Shack 271-217)
R_3, R_{3A}—5,600 ohm, 1/2-watt (Radio Shack 271-031)
$Rect_1$—2-amp, 100-PIV modular bridge rectifier (Radio Shack 276-1171)
T_1, T_{1A}—20-volt to 25.2-volt, up to 500-mA secondary, miniature transformers (Radio Shack 273-1386)

ZENER DIODE-REGULATED SUPPLIES

Circuit 11-10. 5-Volt, Low-Current, Zener-Regulated Supply

Here's a low-cost yet smooth-operating supply suitable for use with many low-current semiconductor circuits. Almost every electronic parts dealer carries the required components, and you probably already have most of them on hand in your spare parts box.

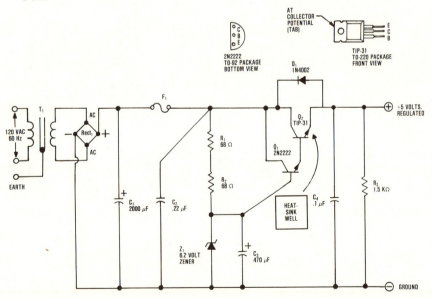

Circuit 11-10. 5-volt, low-current, zener-regulated supply

Under test with a 12.6-volt, 300-mA miniature transformer, this supply easily generated 150 mA into a 32-ohm load (0.74-watts). The voltage drop as the load was attached was 0.2 volts, which calculates to a 4% regulation figure. Ripple to pass transistor Q_2 under full load was measured at 0.4 volts peak-to-peak, with that to the load about 4 mV (a reduction by a factor of 100). Full-load efficiency was 12%, and no-load power consumption from the line was 4-watts.

For assembly pointers, if you plan to use this supply for powering TTL circuitry, make sure that the unloaded output voltage does not exceed 5.25 volts through a tolerance error (maximum TTL voltage), or fall below 4.75 volts because of loading (minimum TTL voltage). If so, substitute another zener diode for Z_1 as these components are subject to about a ± 5% difference in set voltage.

PARTS LIST

C_1—2,000-μF, 25-volt aluminum electrolytic (Radio Shack 272-1020)
C_2—0.22-μF, 50-volt ceramic disk
C_3—470-μF, 16-volt aluminum electrolytic (Radio Shack 272-957)
C_4—0.1-μF, 50-volt ceramic disk (Radio Shack 272-135)
D_1—1N4002 diode (Radio Shack 276-1102)
F_1—Choose to match transformer rating
Q_1—2N2222 or similar transistor (Radio Shack 276-1617)
Q_2—TIP-31 or similar transistor (Radio Shack 276-2017)
R_1, R_2—68 ohm, 1/2-watt (Radio Shack 271-010)
R_3—1,500 ohm, 1/2-watt (Radio Shack 271-025)
$Rect_1$—1-amp, 50-PIV modular bridge rectifier (Radio Shack 276-1161)
T_1—12-volt or 12.6-volt, up to 500-mA secondary, miniature transformer (Radio Shack 273-1385)
Z_1—6.2 volt, 1-watt, zener diode (1N4735) (Radio Shack 276-561)

Circuit 11-11. 9-Volt, Low-Current, Zener-Regulated Supply

For an inexpensive yet very effective answer to your 9-volt, low-current needs, this supply is well worth considering. And you can often cut expenses even further by substituting similar parts that you already have on hand for those which are specified.

As tested with a 12.6-volt, 300-mA miniature transformer this supply easily delivered 150 mA into a 57-ohm load (about 1.3-watts). The output voltage dropped 0.4 volts as the load was attached, which relates to a 4% regulation figure. As for ripple rejection, the 0.7-volt, peak-to-peak waveform to pass transistor

Q_2 was reduced to 10 mV at the load, a reduction by a factor of 70. Efficiency was a moderate 25% under full load, and no-load power draw from the line was 3.9-watts.

For tips, you can trim the output voltage of this supply to near-exact specifications by substituting in various 1N4001 diodes for D_1 and D_2. The forward voltage drop across various diodes ranges from about 0.5 volt to about 1.0 volt, and the output voltage of the supply will vary by a like amount.

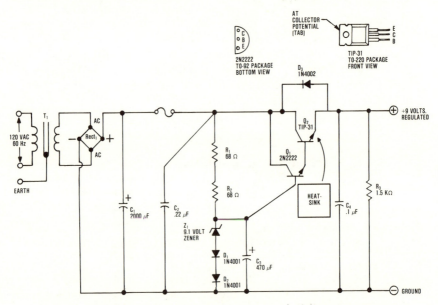

Circuit 11-11. 9-volt, low-current, zener-regulated supply

PARTS LIST

C_1—2,000-μF, 25-volt aluminum electrolytic (Radio Shack 272-1020)
C_2—0.22-μF, 50-volt ceramic disk
C_3—470-μF, 16-volt aluminum electrolytic (Radio Shack 272-957)
C_4—0.1-μF, 50-volt ceramic disk (Radio Shack 272-135)
D_1, D_2—1N4001 diodes (Radio Shack 276-1101)
D_3—1N4002 diode (Radio Shack 276-1102)
F_1—Choose fuse to match transformer rating
Q_1—2N2222 or similar transistor (Radio Shack 276-1617)
Q_2—TIP-31 or similar transistor (Radio Shack 276-2017)
R_1, R_2—68 ohm, ½-watt (Radio Shack 271-010)
R_3—1,500 ohm, ½-watt (Radio Shack 271-025)
$Rect_1$—1-amp, 50-PIV modular bridge rectifier (Radio Shack 276-1161)
T_1—12-volt or 12.6-volt, up to 500 mA secondary, miniature transformer (Radio Shack 273-1385)
Z_1—9.1-volt, 1-watt zener diode (1N4739) (Radio Shack 276-562)

Semiconductor Power Supplies and Auxiliary Circuits 217

Circuit 11-12. 12-Volt, Low-Current, Zener-Regulated Supply

When specifications call for a 12-volt, low-current supply, here's a good cost-effective choice. And because the components are so common, you can probably buy them all in one stop at a parts supply store if you don't already have them left over from previous projects.

When tested with a 12.6-volt, 300-mA miniature transformer, this supply could easily sustain 130 mA into an 87-ohm load (1.5-watts). Voltage drop as the load was attached was 0.6 volts, which works out to a 5% regulation figure. Full-load ripple-to-pass transistor Q_2 was 0.5 volts peak-to-peak, a figure which was reduced to 15 mV at the load for a reduction of 34. Full-load efficiency was a quite good 29%, and no-load power draw was 3.9-watts.

For assembly pointers, you can trim the output voltage to near-exact specifications by substituting various 1N4001 diodes for D_1 and D_2. There is usually some slight difference in voltage drop between diodes, and the output voltage of the supply will change by a like amount.

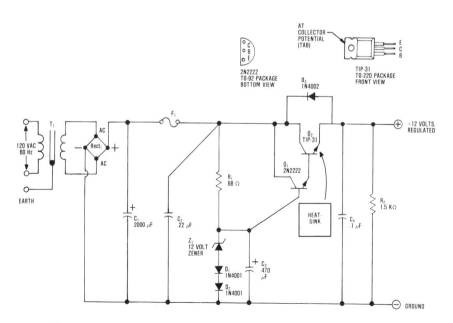

Circuit 11-12. 12-volt, low-current, zener-regulated supply

PARTS LIST

C_1—2,000-µF, 25-volt aluminum electrolytic (Radio Shack 272-1020)
C_2—0.22-µF, 50-volt, ceramic disk
C_3—470-µF, 16-volt, aluminum electrolytic (Radio Shack 272-957)
C_4—0.1-µF, 50-volt ceramic disk (Radio Shack 272-135)
D_1, D_2—1N4001 diodes (Radio Shack 276-1101)
D_3—1N4002 diode (Radio Shack 276-1102)
F_1—Choose to match transformer rating
Q_1—2N2222 or similar transistor (Radio Shack 276-1617)
Q_2—TIP-31 or similar transistor (Radio Shack 276-2017)
R_1—68 ohm, ½-watt (Radio Shack 271-010)
R_2—1,500 ohm, ½-watt (Radio Shack 271-025)
$Rect_1$—1-amp, 50-PIV modular bridge rectifier (Radio Shack 276-1161)
T_1—12-volt or 12.6-volt, up to 500 mA secondary, miniature transformer (Radio Shack 273-1385)
Z_1—12-volt, 1-watt zener diode (1N4742) (Radio Shack 276-563)

Circuit 11-13. 15-Volt, Low-Current, Zener-Regulated Supply

A well-performing 15-volt, low-current supply need not be expensive. Here's one that you can snap together using only standard parts. And because the output voltage is adjustable, you

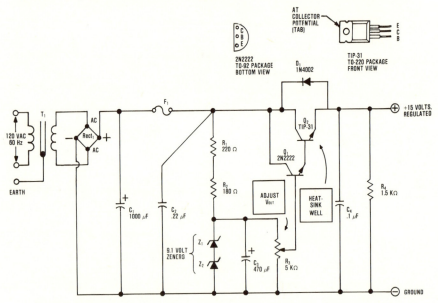

Circuit 11-13. 15-volt, low-current, zener-regulated supply

Semiconductor Power Supplies and Auxiliary Circuits 219

can trim this supply to exact specifications for best performance of driven circuitry.

As tested with a 25-volt, 300-mA miniature transformer, this supply could easily output a continuous 150 mA into a 100-ohm load (2.25-watts). As the load was attached, the output voltage dropped 0.3 volts, which calculates to a 2% regulation figure. Peak-to-peak ripple-to-pass transistor Q_2 under full load was 0.9 volts, a value which was attenuated to 4 mV at the load (a reduction of 225). Full-load efficiency was an adequate 19.5%, and no-load power draw from the line was 7.5-watts.

For tips, be sure to heat-sink pass transistor Q_2 well as it tends to heat under load. If you have a choice of transformers, a 24-volt unit will provide better efficiency though at some slight expense to voltage regulation and ripple rejection. Although this supply is adjustable down to zero volts via "Adjust V_{out}" potentiometer R_3, power dissipation in transistor Q_2 may be excessive at settings below 15 volts when under load.

PARTS LIST

C_1—1,000-μF, 50-volt aluminum electrolytic (Radio Shack 272-1047)
C_2—0.22-μF, 50-volt ceramic disk
C_3—470-μF, 25-volt, aluminum electrolytic (Radio Shack 272-1018 or 272-1030)
C_4—0.1-μF, 50-volt ceramic disk (Radio Shack 272-135)
D_1—1N4002 diode (Radio Shack 276-1102)
F_1—Choose to match transformer rating
Q_1—2N2222 or similar transistor (Radio Shack 276-1617)
Q_2—TIP-31 or similar transistor (Radio Shack 276-2017)
R_1—220 ohm, ½-watt (Radio Shack 271-015)
R_2—180 ohm, ½-watt (Radio Shack 271-014)
R_3—5,000 ohm, ¼-watt, linear-taper potentiometer (Radio Shack 271-217)
R_4—1,500 ohm, ½-watt (Radio Shack 271-025)
$Rect_1$—1-amp, 50-PIV, modular bridge rectifier (Radio Shack 276-1161)
T_1—24-volt or 25.2-volt, up to 500 mA secondary, miniature transformer (Radio Shack 273-1386)
Z_1, Z_2—9.1-volt, 1-watt zener diodes (1N4739) (Radio Shack 276-562)

Circuit 11-14. 1.5-Volt to 9-Volt Adjustable, Low-Current, Zener-Regulated Supply

This versatile supply has dozens of practical uses. Because it will generate better than 150 mA of current anywhere between 1.5 volts and 9 volts, you can use it to produce those non-standard voltages such as 3 volts or 6 volts. You can even use it to power one temporary project, then—with the turn of an adjustment—use it to

220 *Semiconductor Power Supplies and Auxiliary Circuits*

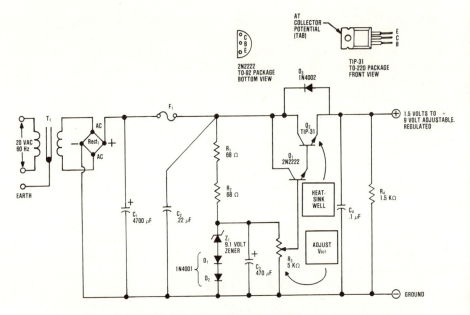

Circuit 11-14. 1.5-volt to 9-volt adjustable, low current, zener-regulated supply

PARTS LIST

C_1—4,700-μF, 25-volt aluminum electrolytic (Radio Shack 272-1022)
C_2—0.22-μF, 50-volt ceramic disk
C_3—470-μF, 16-volt aluminum electrolytic (Radio Shack 272-957)
C_4—0.1-μF, 50-volt ceramic disk (Radio Shack 272-135)
D_1, D_2—1N4001 diodes (Radio Shack 276-1101)
D_3—1N4002 diode (Radio Shack 276-1102)
F_1—Choose to match transformer rating
Q_1—2N2222 or similar transistor (Radio Shack 276-1617)
Q_2—TIP-31 or similar transistor (Radio Shack 276-2017)
R_1, R_2—68 ohm, ½-watt (271-010)
R_3—5,000 ohm, ¼-watt, linear taper potentiometer (Radio Shack 271-217)
R_4—1,500 ohm, ½-watt (Radio Shack 271-025)
$Rect_1$—1-amp, 50-PIV modular bridge rectifier (Radio Shack 276-1161)
T_1—12-volt or 12.6-volt, up to 500 mA secondary, miniature transformer (Radio Shack 273-1385)
Z_1—9.1 volt, 1-watt zener diode (1N4739) (Radio Shack 276-562)

---

power another. Best yet, the cost of assembly can be quite low.

Under test with a 12.6-volt, 300-mA miniature transformer, the load resistance was varied to draw 150 mA as the output voltage was adjusted across its range. Regulation varied from a high of 27% at the 1.5-volt setting, to a low of 4.4% at the 9-volt setting. Peak-to-peak ripple to the load was 1 mV at the 1.5-volt setting, and 2.3 mV at the 9-volt setting. Efficiency was 2.5% at the low setting, and 23.5 at the high setting. No-load power draw from the line was constant at 4 watts.

For practical construction advice, be sure to heat-sink pass transistor Q_2 well as it tends to heat when full current is drawn at the lower voltage settings. To adjust the output voltage, simply rotate "Adjust V_{out}" potentiometer R_3 to the desired setting. Remember to discharge filter capacitor C_1 before you handle it as it can store enough energy to produce a strong spark.

Circuit 11-15. ±9 Volt, Dual Complementary, Low-Current, Zener-Regulated Supply

Here's a well-mannered dual supply suitable for circuits containing op-amps or other components requiring plus/minus voltages. And you can assemble it entirely with inexpensive, widely available parts. Even the transformers are stock, off-the-shelf items.

Under test with dual 12.6-volt, 300-mA transformers, this supply easily delivered 115 mA into a 150-ohm load shunted across the plus/minus output terminals (2-watts). The voltage drop as the load was attached was 0.6 volts, which equates to a 3.4% regulation figure. Under full load, the ripple across the rails feeding pass transistors Q_2 and Q_4 was 1.1 volts peak-to-peak, with that across the load 14 mV (a reduction of 78). Maximum efficiency was about 22%, and no-load power consumption was 7.9-watts.

For some helpful pointers, because this supply consists of symmetrical upper and lower halves which operate in tandem, each corresponding part should match for best performance. We've given these corresponding parts the same number to help you identify them (lower part numbers are followed by an "A" suffix). Be sure to use matched transformers, and to line them up end-to-end to avoid mistakes when interconnecting them. If you wish to trim the output voltages to a close tolerance, try substituting various 1N4001 diodes for D_1/D_2, D_{1A}/D_{2A} as there are slight differences in tolerance.

PARTS LIST

C_1, C_{1A}—2,000-μF, 25-volt aluminum electrolytic (Radio Shack 272-1020)
C_2, C_{2A}—0.22-μF, 50-volt ceramic disk
C_3, C_{3A}—470-μF, 16-volt aluminum electrolytic (Radio Shack 272-957)
C_4, C_{4A}—0.1-μF, 50-volt ceramic disk (Radio Shack 272-135)
D_1, D_{1A}, D_2, D_{2A}, D_3, D_{3A}—1N4001 diodes (Radio Shack 276-1101)

222 *Semiconductor Power Supplies and Auxiliary Circuits*

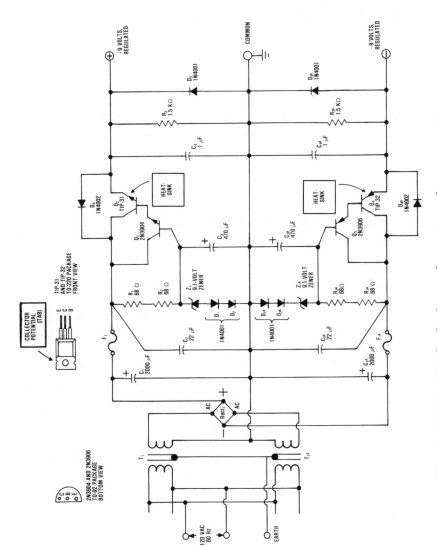

Circuit 11-15. ±9 volt, dual complementary, low-current, zener-regulated supply

Semiconductor Power Supplies and Auxiliary Circuits   223

D_4, D_{4A}—1N4002 diodes (Radio Shack 276-1102)
F_1, F_2—Choose to match transformer ratings
Q_1—2N3904 or similar transistor (Radio Shack 276-2016)
Q_2—TIP-31 or similar transistor (Radio Shack 276-2017)
Q_3—2N3906 or similar transistor (Radio Shack 276-2034)
Q_4—TIP-32 or similar transistor (Radio Shack 276-2025)
R_1, R_{1A}, R_2, R_{2A}—68 ohm, ½-watt (Radio Shack 271-010)
R_3, R_{3A}—1,500 ohm, ½-watt (Radio Shack 271-025)
$Rect_1$—2-amp, 100-PIV modular bridge rectifier (Radio Shack 276-1171)
T_1, T_{1A}—12-volt or 12.6-volt, up to 500 mA secondaries, miniature transformers (Radio Shack 273-1385)
Z_1, Z_{1A}—9.1-volt, 1-watt zener diodes (1N4739) (Radio Shack 276-562)

Circuit 11-16. ±15 Volt, Dual Complementary, Low-Current, Zener-Regulated Supply

When best performance of op-amp and other circuitry calls for a plus/minus 15-volt supply, the following diagram (Circuit 11-16) can be the low-cost, easy-to-build solution. Adjustable outputs allow you to fine-tune each output voltage to exact specifications ... and 100% standard-part composition usually means one-stop shopping.

Under test with twin 25.2-volt, 300-mA miniature transformers, this supply easily generated 100 mA into a 300-ohm load shunted across the output terminals (3-watts). The output voltage dropped 0.5 volts as the load was attached for a regulation figure of 1.7%. Peak-to-peak ripple across the rails feeding the pass transistors was 1.5 volts, with that to the load a mere 6 mV (a reduction of 250). Full-load efficiency was about 14%, and no-load power drain was about 15-watts.

Useful tips to help you construct this supply: Since this unit consists of symmetrical upper and lower halves which operate in tandem, each corresponding upper and lower part should match for best performance. We've given these corresponding parts the same number to help you identify them (lower part numbers are followed by an "A" suffix). Be sure to use matched transformers ... and to line them up end-to-end so that you can't make a mistake when interconnecting them. To adjust the plus/minus outputs for 15 volts with respect to common, simply rotate "Adjust V_{out}" potentiometers R_3/R_{3A}.

224 Semiconductor Power Supplies and Auxiliary Circuits

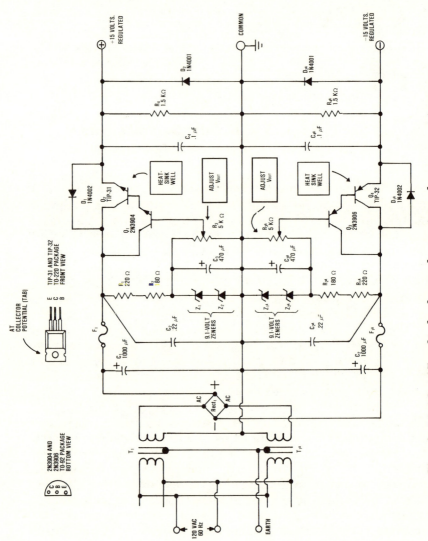

Circuit 11-16. ±15 volt, dual complementary, low-current, zener-regulated supply

PARTS LIST

C_1, C_{1A}—1,000-μF, 50-volt aluminum electrolytic (Radio Shack 272-1047)
C_2, C_{2A}—0.22-μF, 50-volt ceramic disk
C_3, C_{3A}—470-μF, 25-volt aluminum electrolytic (Radio Shack 272-1018 or 272-1030)
C_4, C_{4A}—0.1-μF, 50-volt ceramic disk (Radio Shack 272-135)
D_1, D_{1A}—1N4002 diodes (Radio Shack 276-1102)
D_2, D_{2A}—1N4001 diodes (Radio Shack 276-1101)
F_1, F_{1A}—Choose to match transformer ratings
Q_1—2N3904 or similar transistor (Radio Shack 276-2016)
Q_2—TIP-31 or similar transistor (Radio Shack 276-2017)
Q_3—2N3906 or similar transistor (Radio Shack 276-2034)
Q_4—TIP-32 or similar transistor (Radio Shack 276-2025)
R_1, R_{1A}—220 ohm, ½-watt (Radio Shack 271-015)
R_2, R_{2A}—180 ohm, ½-watt (Radio Shack 271-014)
R_3, R_{3A}—5,000 ohm, ¼-watt, linear taper potentiometer (Radio Shack 271-217)
R_4, R_{4A}—1,500 ohm, ½-watt (Radio Shack 271-025)
$Rect_1$—2-amp, 200-PIV modular bridge rectifier (Radio Shack 276-1173)
T_1, T_{1A}—24-volt or 25.2 volt, up to 500 mA secondary, miniature transformers (Radio Shack 273-1386)
Z_1, Z_{1A}, Z_2, Z_{2A}—9.1-volt, 1-watt zener diodes (1N4739) (Radio Shack 276-562)

Circuit 11-17. 5.25-Volt, Battery-Powered Supply for TTL Circuitry

For portable or short-term operation of TTL/LS-based and other circuitry, this simple battery supply can be immensely practical. Diode D_1 drops battery voltage to the correct 5.25-volt level, and resistor R_1 in conjunction with capacitor C_1 suppresses transients which might otherwise cause false triggering of the driven circuitry.

For assembly tips, to fit a confined space use smaller batteries such as the "AAA" or "AA" type. Otherwise, larger batteries such as the "C" or "D" type will usually give longer service. Mount the cells in a series-connected battery holder, a device which is available for most cell types at almost all electronic supply stores.

PARTS LIST

B_1—B_4—Four 1½-volt dry cell batteries
C_1—10-μF, 25-volt solid tantalum (optional) (Radio Shack 272-1423 is OK)
D_1—1N4001 diode (Radio Shack 276-1101)
R_1—0.56 ohm, ¼-watt or more (optional) (Radio Shack 271-072)
S_1—ON/OFF switch (Radio Shack 275-602)

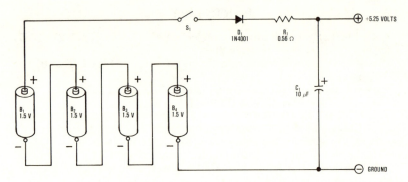

Circuit 11-17. 5.25-volt battery-powered supply for TTL circuitry

Circuit 11-18. 9-Volt, Low-Current Battery-Powered Supply

Here's a good 9-volt unit for energizing low-current circuitry. Use it anywhere that space, portability, or convenience requirements make a line-powered supply impractical—it's especially handy for breadboarding. Filter components C_1 and R_1 dampen transients which might otherwise cause instability in the driven circuitry, and switch S_1 disconnects the supply to conserve energy.

For construction pointers, a battery clip of the type used in many transistor radios is usually the easiest way to make connection with the battery (available at most electronic supply stores). An alkaline battery offers longer shelf life and higher

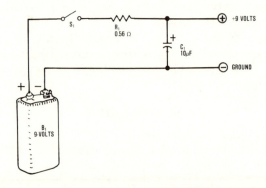

Circuit 11-18. 9-volt, low-current, battery-powered supply

Semiconductor Power Supplies and Auxiliary Circuits 227

current capabilities than a conventional carbon/zinc type, albeit at a correspondingly higher price. For longer battery service or to meet higher current demands, shunt together two or more batteries of the same type.

PARTS LIST

B_1—Any 9-volt, transistor radio-type battery
C_1—10-μF, 25-volt solid tantalum (optional) (Radio Shack 272-1423 is OK)
R_1—0.56 ohm, ¼-watt or more (optional) (Radio Shack 271-072)
S_1—ON/OFF switch (Radio Shack 275-612)

Circuit 11-19. ±9 Volt, Dual Complementary, Low-Current, Battery-Powered Supply

A plus/minus supply for portable or short-term use can be as simple as two 9-volt batteries and eight inexpensive parts. Switch S_1 and S_2 may even be omitted if this supply is to be used for breadboarding. Although this unit is compact enough to be housed in an enclosure the size of a package of cigarettes, it provides good transient suppression characteristics.

For construction tips, the easiest way to connect to the batteries is with a clip of the type used in most transistor radios. For longer battery service or to meet higher current demands, you can shunt each battery with an additional cell of the same type.

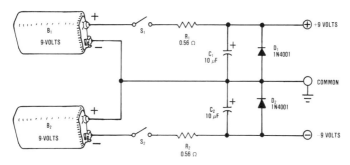

Circuit 11-19. ±9 volt, dual complementary, low-current, battery-powered supply

PARTS LIST

B_1, B_2—9-volt, transistor radio-type batteries
C_1, C_2—10-μF, 25-volt, solid tantalum (optional) (Radio Shack 272-1423 are OK)
D_1, D_2—1N4001 diodes (Radio Shack 276-1101)
R_1, R_2—0.56 ohm, ¼-watt or more (optional) (Radio Shack 271-072)
S_1, S_2—ON/OFF switches (optional) (Radio Shack 275-612)

Circuit 11-20. 12-Volt, High-Current Battery-Powered Supply

For 12-volt, high-current applications or for continuous circuit use, an automotive-type lead-acid battery can make a very cost-effective supply. Plenty of power for operating lights, relays, and sounders! As an added plus, this type of supply is completely rechargeable, can last for years, and is immune to power outages (important in alarms and other circuits).

As assembly pointers, you can use a motorcycle-type battery where weight or space consumption is a factor ... an automotive-type battery for longer service between recharges. Lead-acid batteries are, of course, power-rated by amp-hours. A 120 amp-hour battery, for example, should be able to output 1 amp for 120 hours; ½ amp for 240 hours (etc.). A hydrometer, available at automotive supply stores, is a handy tool for testing this type of battery to determine its state of charge.

Try to position this supply as close as possible to the circuitry that it will drive. If the intervening distance is more than a few feet, mount transient-suppressing components R_1 and C_1 in close proximity to the driven circuitry to dampen noise picked up by the interconnecting wires (they act as an antenna). If more than 1 amp of current is to be drawn from the battery at any time, omit resistor R_1.

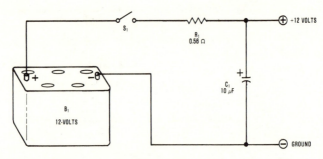

Circuit 11-20. 12-volt, high current, battery-powered supply

PARTS LIST

B_1—Any 12-volt, lead-acid battery
C_1—10-μF, 25-volt solid tantalum (optional) (Radio Shack 272-1423)
R_1—0.56 ohm, 1-watt (optional) (Radio Shack 271-027)
S_1—ON/OFF switch (Radio Shack 275-324 is OK for most applications)

DECOUPLER/REGULATOR FOR USE WITH MOBILE PROJECTS

Circuit 11-21. Decoupler/Regulator for Use with Mobile Projects

Although an automotive-type battery is a good source of power for use with the more delicate varieties of semiconductor circuitry, all bets are off when this same battery is simultaneously used in a motor vehicle. Voltage spikes, which are generated by a variety of causes and regularly pass through the wiring system, are easily capable of causing instability in, if not physical damage to, sensitive semiconductor components. What's more, using a 24-volt booster to start the engine or connecting jumper cables improperly usually means instant destruction for such frail parts.

Protect the delicate semiconductor components of your mobile projects from almost certain harm with the decoupler/regulator illustrated in the following diagram (Circuit 11-21). The 5-volt regulated output will accommodate most types of semiconductor circuits, and will provide them with up to 100 mA of current. The 10-volt output can accommodate most of the rest, although with only fair regulation.

For construction, install this protective circuit inside the same metallic circuit box used for the project circuitry (if possible). Be sure to mount the box in a cool, protected location, and ground it to the vehicle's chassis. To wire this circuit into the vehicle's system, first attach the negative power lead as close as possible to, if not directly on, the negative pole of the battery. This will minimize pickup of unwanted transient voltages. Next, connect the positive power lead to an always-hot point (such as the clock fuse or positive pole of the battery) if circuitry must operate with the ignition switch OFF ... otherwise to a fuse feeding relative transient-free gear such as the radio if circuitry must operate while the ignition switch is ON.

230   *Semiconductor Power Supplies and Auxiliary Circuits*

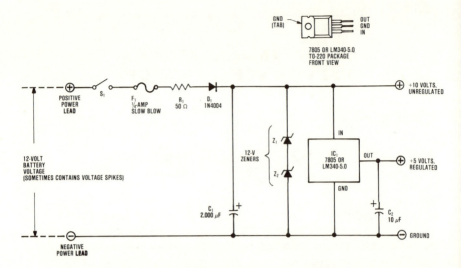

Circuit 11-21. Decoupler/regulator for use with mobile projects

PARTS LIST

C_1—2,000-μF, 25-volt aluminum electrolytic (Radio Shack 272-1020)
C_2—10-μF, 25-volt solid tantalum
D_1—1N4004 diode (Radio Shack 276-1103)
F_1—¼-amp, slow blow fuse
IC_1—7805 or LM340-5.0 3-terminal voltage regulator (Radio Shack 276-1770)
R_1—50 ohm, 1-watt (Radio Shack 271-133)
S_1—ON/OFF switch (Radio shack 275-612)
Z_1, Z_2—12-volt, 1-watt zener diodes (1N4742) (Radio Shack 276-563)

Wrapping up, power supply construction is no place for shortcuts. Quite aside from safety considerations, many if not most breakdowns in electronic gear are directly traceable to a malfunctioning power supply, which in turn is often caused by undersized (or too few) components. For these reasons, all line-powered supplies in this chapter are designed to deliver that rugged, long-term service you will need. In most cases they should stand up to 5 full years of non-stop operation.

INDEX

A

Adhesive-backed copper foil, 34
Adjustment specifications, 41
Air conditioning, 58
Alarm, intrusion, 135, 145-146 (see also Warning mechanism)
Alarm ON Time Adjust R_4, 180
Alarm ON Time Adjust, R_6, 174
Alarm system, 173-176 (see also Warning mechanism)
Ambient Light Adjust, 53, 54
Ambient Light Controller, 51, 60-62
Amplifier, stereo, 31-41, 117, 124-126
Application, 20, 21
Arm Switch S_1, 176
Assembly of circuit, 31-41 (see also Circuit)
Attenuate Adjust R_1 and R_4, 131
Audio-signal generator, 192-193
Automatic pump control, 72-74
Automotive-type battery, 41

B

Back-up sounder, 135, 136-137
Basement floods, 72-74
Battery:
  charger, 135, 138-140
  refresher, 71-72
  symbol, 41
Battery back-up unit, 163, 164
Battery-powered supply:
  5.25-volt, for TTL circuitry, 225-226
  9-volt, low-current, 226-227
  ±9 volt, dual complementary, low-current, 227
  12-volt, high-current, 228
Battle lantern, 65, 66-67
Beacon light, 81, 83-84
Beat generator, 152-153
Bell, symbol, 41
Bi-polar transistors, 189
"Black light," 77
Blender, 58
Brake fluid monitor, 135, 140-141
Breadboarding:
  defective or weak parts, 25
  end in itself, 25
  first, 24
  heavy copper wire, 25

Breadboarding (cont.)
  mount components, 26
  painting board white, 26
  preparing socket, 26
  pre-test circuit, 25
Break-beam, 164, 172-173
Buzz Box, 185, 189-190
Buzzer, 41, 136, 165, 166, 172, 173

C

Capacitance, value, 193
Capacitance Box, 185, 187-188
Capacitance Tester, 185, 193-195
Capacitor:
  substituting, 29, 30
  symbols, 42
  testing, 189
  unmarked, 185, 187
Capacitor$_3$, 173
Car:
  accidents when backing up, 135
  back-up sounder, 135, 136-137
  Charge Level R_3, 138
  cold-weather, 135
  decoupler/regulator, 135, 140, 141, 143
  dependable intrusion alarm, 135, 145-146
  headlamp-ON reminder, 135, 143-144
  idiot light audio back-up, 135, 141-142
  low brake fluid monitor, 135, 140-141
  piezo buzzer, 136
  problem, 135
  protect circuit from damage, 135
  safety and convenience, 135
  Sensitivity Adjust R_2, 141
  Switch S_1, 140, 141, 145
  Switches S_2, S_3 ... S_x, 145
  Tone Adjust R_4, 136
  12-volt automotive battery charger, 135, 138-140
  12-volt to 120-volt inverter, 137-138
Carbon-zinc battery:
  refresher, 71-72
  symbol, 41

232 *Index*

Ceramic disk capacitor, 42
Ceramic phonograph cartridge, 42
Chafe guard, 28
Charge Level R_3, 138
Chemicals cabinet, 164-166
Circuit:
  assembling, 31-41
    adhesive-backed copper foil, 34
    adjustment specifications, 41
    component values, 41
    face of enclosure box, 36
    fuse sizes, 41
    gathering components, 32
    holes drilled, 38-39, 41
    input/output devices, 38
    label controls, 38
    mount components, 33-34
    necessary interconnections, 38
    nip off excess lead lengths, 35-36
    ohmmeter, 38
    parts list, 32
    potentiometers, 38
    pre-tinned bus-type wire, 34
    printed circuit techniques, 34
    safety glasses, 35
    secure perfboard, 36-38
    silicon grease, 32-33
    sketch, 41
    soldering, 35
    underside of board, 34
    wire-wrap techniques, 34
  integrated (*see* Integrated circuits)
  schematic symbols, 41-47 (*see also* Symbols)
Circulating pump, 81
Coarse Adjust R_1, 153
Coarse ON Time Adjust R_5, 101
Coding system, 20, 21
"Cold" wire, 28
Collectors, solar, 81
Comfort, 22, 23
Commercial Killer, T.V., 65, 70-71
Components:
  adequate cooling, 28-29, 38-39, 41
  gather, 32
  mount on perfboard, 29, 33-34
  obtaining, 20-21
  permanent mounting, 26
  temporary mounting, 26
  values, 41
Connectors, plastic-insulated, 27
Costs, 20, 21, 29-30
Current, preset amount, 185, 188
Current Adjust R_2, 154, 155

D

Data books, 22
Data sheet, 21
Dawn tasks, 99, 105-107
Decals, 38
Decoupler/regulator, 135, 140, 141, 143, 229-230
De-Magnetizer, 149, 151-152
Dependability, wire for, 30-31
Dimmer/motor speed control, 52-53
Diodes:
  discrete, 44
  generic number, 20
  light-emitting, 43
  symbols, 42-43
  testing, 189
  zener, 20, 29, 42
Discrete transistors, 20
Doorbell of Two Tones, 65, 67-68
Drapes, opening, 86, 88
Dry cell battery:
  refresher, 71-72
  symbol, 41
Dual-throw switch S_1, 151
Dusk or dawn tasks, 99, 105-107

E

"Earth" wire, 28
Electric Flypaper, 65, 76-77
Electrolytic capacitors, 29, 42
Electroplating Machine, 149, 153-155
Electronic Scarecrow, 65, 74-76
Electronic Switch With Memory, 99, 101-103
Enclosure box:
  chafe guard, 28
  high voltage short circuit, 27
  lay-out the face, 36
  line connections inside, 27
  secure perfboard, 36-38
  where line cord enters, 28
Energy efficiency in home, 65 (*see also* Home)
Energy savers:
  Ambient Light Adjust, 53, 54
  Ambient Light Controller, 51, 60-62
  blender, 58
  dimmer/motor speed control, 52-53
  exhaust fan, 51, 58
  fuel bill, 55-56

Index

Energy savers (*cont.*)
  garage light, 58
  heating and air conditioning, 58
  heat lamp, 58
  incandescent lamp, 52
  light miser, 53-54
  power factor correction, 51, 54-55
  pushbutton switch that "remembers," 58-60
  Temperature Limit Adjust R_4, 58
  Temperature Sensing Fan Regulator, 51, 56-58
  thermostat, 55-56
  Triac, 51, 52, 54
  turn OFF appliances, 51
  ventilation fan, 51, 58
Escape route, 65
Exhaust fan, 51, 58

F

Fan:
  exhaust, 51, 58
  ventilation, 51, 58
Field-effect transistor, symbol, 47
Fine Adjust R_2, 153
Fine ON Time Adjust R_4, 101
Fire, 65, 66
Fire alarm, 167, 168
5-volt, high-current, 202-204
5-volt, low-current:
  IC-regulated supply, 200-201
  zener-regulated supply, 214-215
5-volt, medium-current, 201-202
5.25-volt, battery powered supply, 225-228
15-volt, low-current:
  IC-regulated supply, 207-208
  zener-regulated supply, 218-219
±15 volt, dual complementary, low-current, 223-225
±15 volts to ±20 volts, dual complementary, 212-214
Flasher, 81, 83-84
Flash Interval variable resistor R_2, 84
Floods, 66, 72
Flypaper, Electric, 65, 76-77
Frequency Adjust R_2, 191
Frequency response, 191-192
Fuel bill, 55-56
Fun and profit:
  amusement at parties, 149
  Beat generator, 152-153
  charging screwdrivers, 149

Fun and profit (*cont.*)
  charging tack hammers, 149
  Course Adjust R_1, 153
  Current Adjust R_2, 154, 155
  CW oscillator, 149, 150-151
  dual-throw switch S_1, 151
  Electroplating Machine, 149, 153-155
  Fine Adjust R_2, 153
  Gain Adjust R_5, 158, 159
  Lie Detector, 149, 155-156
  listen for defects, 149
  Magnetizer/De-Magnetizer, 149, 151-152
  Morse Code, 149, 150
  neutralize unwanted charge, 149
  pick up inaudible vibrations, 149
  Press to Reset switch S_1, 153
  Rate Adjust R_1, 156, 158
  reduce audio noise, 149
  restore high-frequency performance, 149
  rotary switch S_1, 156
  scrap parts, 149
  Sound Microscope, 149, 158-159
  Sports Starting Light Sequencer, 156-158
  Standby/Countdown Switch$_1$, 156, 158
  telegraph key contacts, 149
  Tone Adjust R_1, 151
  "Vernier" variable resistor R_3, 153
  Volume Adjust R_4, 151
  Volume Adjust R_8, 158, 159
Fuse, 30, 41, 43
Fuse box, 65
Fuse holder, 43

G

Gain Adjust R_3 and R_{3A}, 126
Gain Adjust R_5, 158, 159
Gain Adjust R_5 and R_{5A}, 128
Garage light, 58
Garden pests, 65, 74-76
Generator, Solar Electric, 81, 82-83, 89
Generic system:
  diodes, 20
  discrete transistors, 20
  grade of device, 20
  IC MASTER, 21
  industry-wide, 20
  integrated circuits, 20
  manufacturer, 20

Generic system (cont.)
  "prefix" and "suffix" codes, 20, 21
  GO light sequencer, 156-158
  Grade of device, 20

H

Headlamp-ON reminder, 135, 143-144
Heat:
  circuit components, 28-29, 38-39, 41
  solar energy, 81
Heater, water, 100-101
Heating system, 55-56, 58
Heat lamp, 58
Hi-Fi Magnetic Cartridge Preamp, 117, 128-129
Hi/Low Logic Tester, 185-186
High-temperature alarm, 164, 167-169
High voltage:
  chafe guard, 28
  connections inside enclosure box, 27
  "earth" wire, 28
  plastic-insulated connectors, 27
  safety, 27-28
  source of leakage or shock, 28
  third-wire ground 28
  three-pronged plug, 28
High-water alarm, 164, 166-167
Home:
  automatic pump control unit, 72-74
  Battle Lantern, 65, 66-67
  carbon-zinc battery refresher, 71-72
  Doorbell of Two Tones, 65, 67-68
  Electric Flypaper, 65, 76-77
  Electronic Scarecrow, 65, 74-76
  energy efficiency, 65
  escape route, 65
  fuse box, 65
  low cost of semiconductors, 65
  power can fail, 65, 66
  Rain Alarm, 65, 68-69
  safety and convenience, 65
  silent, hidden servant, 65
  T.V. Commercial Killer, 65, 70-71
Hurricanes, 66

I

Idiot light audio back-up, 135, 141-142
Incandescent light bulb, 43, 52, 83
Information on IC's, 21-22
Inhibit Relay 2, 3, 4, etc. 113

Inhibit Switches, 111, 113
Inline Current Limiter, 185, 188-189
Input/output devices, 38
Insects, flying, 65, 76-77
Integrated circuit:
  access to information, 21-22
  assembling, 31-41 (see also Circuit)
  attitude of technician, 19
  breadboarding, 26
  circuit design handbooks, 22
  data books, 22
  data libraries, 22
  designing own, 22
  entire block functions, 19
  free application notes, 21
  free data sheet, 21
  generic number, 20, 21
  grade of device, 20
  IC MASTER, 21
  impact on society, 19
  key electric parameters, 21
  low cost, 19
  manufacturer, 20, 21
  National Semiconductor, 22
  photographic process, 19
  "prefix" and "suffix" codes, 20, 21
  Radio Shack, 20
  reference manuals, 22
  70's, 19
  small size, 19
  stereo amplifier, 31-41
  symbol, 43
  Texas Instruments, Inc., 22
  voltage regulators, 30
Integrated circuit-regulated supplies:
  5-volt, high-current, 202-204
  5-volt, low-current, 200-201
  5-volt, medium-current, 201-202
  9-volt to 12-volt adjustable, low-current, 204-205
  9-volt to 12-volt adjustable, medium-current, 205-207
  ±9 volt to ±12 volt, dual complementary, 210-211
  12-volt to 15-volt adjustable, high-current, 208-210
  15-volt, low-current, 207-208
  ±15 volts to ±20 volts, dual complementary, 212-214
Intercom, 117, 119-120
Intrusion alarm, car, 135, 145-146

K

Kill Sound, 70, 71

Index

L

Labels, 38
Lamp:
  heat, 58
  incandescent, 52
Lantern-type battery, 41
Lead-acid cells, 83
Leakage, source, 28
Left-to-Right Sun Tracker, 81, 89-91
Letter-punching tool, 38
Lie Detector, 149, 155-156
Light bulb, symbol, 43
Light detector, 178-179
Light-emitting diode, 43
Lighting, good, 22, 23
Light Level Adjust, 88, 89, 93, 95
Light miser, 53-54
Lights:
  failure, 66
  Self-Powered Strobe Light/Beacon Flasher, 81, 83-84
Light sequencer, 156-158
Limit Switch, 91
Line-powered alarm, 163, 164
Line-powered circuit, 27
Lock Switch S_6, 174
Low brake fluid monitor, 135, 140-141
Low-temperature alarm, 164, 169-171

M

Magazines, 20
Magnetizer/De-Magnetizer, 149, 151-152
Manuals, 22
Manufacturer, 20, 21
Measurements (see Testing and measurements)
Medicine cabinet, 163, 164-166
Memory, robot, 99, 101-103
Microphone, symbol, 44
Mobile projects, 229-230
Modular bridge rectifiers, 29, 44
Morse Code, 149, 150
Motor, symbol, 43
Mounting components, 29, 33-34
Multi-cell battery, 41
Multimeters, 185

N

National Semiconductor, 22
Negative Probe, 103
Neutral wire, 28
Nickel-cadmium cells, 83

9-volt, low-current:
  battery powered supply, 226-227
  zener-regulated supply, 215-216
9-volt to 12-volt adjustable:
  low-current, 204-205
  medium-current, 205-207
±9 volt, dual complementary, low current:
  battery-powered supply, 227
  zener-regulated supply, 221-223
±9 volt to ±12 volt, dual complementary, 210-211
NPN transistor, 47

O

Ocean surf, sound, 117, 118
Ohmmeter, 38, 189
1.5-volt to 9-volt adjustable, low current, 219-221
ON/OFF switch, symbol, 46
ON Time Adjust, 88, 89, 93, 95
ON Time Adjust R_5, 111
ON Time Adjust R_8, 111
ON Time Adjust R_9, 109
ON Time Adjust R_{11}, 105
Opens, 185, 186, 189
Orchard pests, 65, 74-76
Oscillator, CW, 149, 150-151
Oscilloscope, 23, 24, 185, 191, 192

P

PA/loudhailer system, 117
PA/Megaphone, 122-124
Parts list, 32
Perfboard:
  mount components, 33-34
  secure to enclosure box, 36, 38
Pests:
  garden and orchard, 65, 74-76
  insects, 65, 76-77
Phase cancellation, 45
Phonograph cartridge, ceramic, 42
Photo-cell R_1, 178
Photo-cell R_2, 172
Photoresistor, 86, 88, 89, 93, 95
Piezo buzzer, 136, 173
"Pink noise" generator, 117, 118-119
Plug, three-pronged, 28
PNP transistor, symbol, 47
Popular Electronics, 20
Potentiometers, 38, 45
Power cables, 30
Power failure, 65, 66
Power factor correction, 51, 54-55

Index

Power resistors, 29
Power supplies:
  battery, 225-228
  bench-type regulated, 24
  decoupler/regulator, 229-230
  IC-regulated, 200-214
  mobile projects, 229-230
  TTL circuitry, 225-228
  warning mechanisms, 163, 164
  zener diode-regulated, 214-225
Preamp, 117, 126-127, 128-129
Prefix and suffix codes, 20, 21
Press to Begin Sequence switch S_2, 111
Press to Reset switch S_1, 153
Press to Test switch S_2, 193
Pre-tinned bus-type wire, 34
Profit (see Fun and profit)
Pump, circulating, 81
Pump control unit, 72-74, 81, 84-86
Pushbutton switch that "remembers," 58-60
Push to Reload switch S_1, 180

R

Radio Shack stores, 20
Rain Alarm, 65, 68-69
Rain-on-roof, sound, 117, 118
Range Adjust switch S_1, 191
Rate Adjust R_1, 156, 158
Reference manuals, 22
Relay, symbol, 44
Relay Controller Which "Hears," 107-109
Relay Controller Which "Sees," 99, 105-107
Relay$_1$, 109, 111, 113
Relay$_2$, 111, 113
Relay$_3$, 113
Relay$_4$, 113
Relay$_5$, 111
Relay$_8$, 111
Relay ON Time Adjust R_4, R_{10}, etc., 113
Resistor R_8, 193
Resistors, symbols, 44-45
Return-To-Dawn, 89, 91-93
Robot circuits:
  adjustable timing range, 99
  Coarse ON Time Adjust R_5, 101
  cost, 99
  discrete steps in sequence, 99
  dusk or dawn tasks, 99, 105-107
  Electronic Switch with Memory, 99, 101-103
  fairly complex jobs, 99

Robot circuits (cont.)
  Fine ON Time Adjust R_4, 101
  freeing time, 99
  Inhibit Relay 2, 3, 4, etc., 113
  Inhibit Switches, 111, 113
  "memory" programmed, 99
  Negative Probe, 103
  1-2-3 Robot, 99
  ON Time Adjust R_5, 111
  ON Time Adjust R_8, 111
  ON Time Adjust R_9, 109
  ON Time Adjust R_{11}, 105
  precise-temperature water heater, 100-101
  Press to Begin Sequence switch S_2, 111
  Relay Controller which "Hears," 107-109
  Relay Controller which "Sees," 99, 105-107
  Relay 1, 109, 111, 113
  Relay 2, 111, 113
  Relay 3, 113
  Relay 4, 113
  Relay 5, 111
  Relay 8, 111
  Relay ON Time Adjust R_4, R_{10}, etc., 113
  routine tasks, 99
  Sensitivity Adjust R_2, 109
  Sensitivity Adjust R_3, 103
  Sensitivity Adjust R_4, 103
  Sensor Probes, 103
  Standby/Start Sequence switch S_1, 113
  Standby/Timed Interval switch, 101
  Tank Filling Robot, 99, 103-104
  Temperature Adjust R_3, 100-101
  time intervals, 99
  Tireless 1-2-1-2 Robot, 99, 109-111
  Trigger Level Adjust R_3, 105
  turning OFF appliance, 99
  two-step function, 99
Rotary switch, 46
Rotary switch S_1, 156

S

Safety, 27-28
Safety glasses, 35
Scarecrow, Electronic, 65, 74-76
Scrap chassis, 21
Scrap parts, 149
Sensitivity Adjust R_1, 176, 177
Sensitivity Adjust R_2, 69, 109, 141, 167

Index 237

Sensitivity Adjust R_3, 74, 103, 178, 179
Sensitivity Adjust R_4, 74, 103
Sensor Probes, 72, 74, 103
Series Resistance Select switch S_1, 193
Servants, silent, 65
Shock, source, 28
Shorts, 65, 185, 186, 188, 189
Signal-to-noise ratio, 117
Silent sentries, 163-164
Silicon, overheated, 185
Silicon-Controlled Rectifier, symbol, 45
Silicon grease, 32-33
Silicon solar cell, symbol, 45
Sine waves, 185
Single-pole, dual-throw switch, 46
Single-pole, single-throw switch, 46
Solar energy:
  circulating pump, 81
  collectors, 81
  fairly concentrated, 81
  Flash Interval variable resistor R_2, 84
  free, 81
  gathering its heat, 81
  inexhaustible, 81
  Left-to-Right Sun Tracker, 81, 89-91
  Light Level Adjust, 88, 89, 93, 95
  Limit Switch, 91
  On Time Adjust, 88, 89, 93, 95
  opening drapes, 86, 88
  photoresistor, 86, 88, 89, 93, 95
  pump control, 81, 84-86
  Return-To-Dawn, 89, 91-93
  Self-Powered Strobe Light/Beacon Flasher, 81, 83-84
  solar array and collection tank, 81
  solar cells, 81, 82
  Solar Electric Generator, 81, 82-83, 89
  storage cells, 82-83
  Temperature Offset Adjust R_2, 84, 86
  Thermistors R_3 and R_6, 86
  trap, 86-89
  ups/down control for solar array, 93-95
  your locality, 81
Soldering, 35
Sound:
  Attenuate Adjust R_1 and R_4, 131
  boost unit for audio set, 117
  Gain Adjust R_3 and R_{3A}, 126
  Gain Adjust R_5 and R_{5A}, 128

Sound (*cont.*)
  Hi-Fi magnetic Cartridge Preamp, 117, 128-129
  low-powered audio system, 120-121
  ocean surf, 117, 118
  PA/loudhailer system, 117
  PA/Megaphone, 122-124
  "pink noise" generator, 117, 118-119
  professional unit at affordable price, 117, 130-131
  rain-on-roof, 117, 118
  reduce distortion in main amplifier, 117
  signal-to-noise ratio, 117
  SPDT momentary-contact press-to-talk switches, 120
  Stereo/Mono switch, 130, 131
  stereo phonograph/tape recorder amplifier, 117, 124-126
  Super Stereo Preamp, 117, 126-127
  Tone Adjust R_4, 124
  Tone Adjust R_5, 124
  2-Way Intercom, 117, 119-120
  Volume Adjust R_2, 118, 122, 124
  Volume Adjust R_3, 121, 124
  wind-through-trees, 117
Sound Microscope, 149, 158-159
Spark Rate Adjust variable resistor R_3, 77
SPDT momentary-contact press-to-talk switches, 120
Speaker, 45, 74, 76, 190-192
Square waves, 185
Sports Starting Light Sequencer, 156-158
Standby/Countdown Switch$_1$, 156, 158
Standby/Kill Sound switch S_1, 70
Standby/Start Sequence switch S_1, 113
Standby/Timed interval switch, 101
Stereo amplifier, 31-41, 117, 124-126
Stereo/Mono switch, 130, 131
Stereo Preamp, 117, 126-127, 128-129
Storage cells, 82-83
Strobe light, 81, 83-84
Sump, 72
Switches, symbols, 47
Switches Sa through Sd, 173
Switches S_2, S_3 ... S_x, 145, 173
Switch S_1, 140, 141, 145, 165, 173
Switch S_5, 174
Symbols:
  battery, 41
  buzzer or bell, 41
  capacitors, 42

238

Index

Symbols (cont.)
ceramic phonograph cartridge, 42
diode, 42-43
fuse holder and fuse, 43
integrated circuit, 43
light bulb, 43
microphone, 44
modular bridge rectifier, 44
motor, 43
photoresistor, 44
relay, 44
resistors, 44-45
Silicon-Controlled Rectifier, 45
silicon solar cell, 45
speaker, 45
switch, 46
thermistor, 46
transformer, 47
transistors, 47
Triac, 47

T

Tank Filling Robot, 99, 103-104
Telegraph key contacts, 149
Temperature:
  high, 164, 167-169
  low, 164, 169-171
Temperature Adjust R_1, 168
Temperature Adjust R_3, 100-101, 167, 168, 169, 171
Temperature Light Adjust R_4, 58
Temperature Offset Adjust R_2, 84, 86
Temperature Sensing Fan Regulator, 51, 56-58
Testing and measurements:
  audio-signal generator, 192-193
  Buzz Box, 185, 189-190
  Capacitance Box, 185, 187-188
  Capacitance Tester, 185, 193-195
  capacitors (out of circuit), 189
  continuity/approximate resistance, 189
  decrease overhead cost, 185
  designing circuits, 185, 187
  diodes and bi-polar transistors, 189
  discrete components, 185
  Frequency Adjust R_2, 191
  frequency response, 191-192
  Hi/Low Logic Tester, 185, 186
  increase productivity, 185
  Inline Current Limiter, 185, 188-189
  keep parts count low, 185, 186
  multimeters, 185
  opens, 185, 186, 189

Testing and measurements (cont.)
  oscilloscopes, 185, 191, 192
  overheated silicon, 185
  preset amount of current, 185, 188
  Press to Test switch S_2, 193
  Range Adjust switch S_1, 191
  Resistor R_8, 193
  Series Resistance Select switch S_1, 193
  shorts, 185, 186, 188, 189
  sine waves and square waves, 185
  transistor Q_1, 188
  unmarked capacitor, 185, 187
  value of capacitance, 193
  variable-frequency square-wave generator, 190-192
  Volume Adjust R_3, 189
Test instruments, 24
Texas Instruments Incorporated, 22
Texts, 22
Thermistor, symbol, 46
Thermistor R_1, 167, 168
Thermistor R_2, 169, 171
Thermistors R_3 and R_6, 86
Thermostat, 55-56
Third-wire ground, 28
Three-pronged plug, 28
Time Adjust variable resistor R, 70
Time Interval Adjust R_1, 76
Timer, 179-181
Tone Adjust R_1, 151
Tone Adjust R_4, 124, 136
Tone Adjust R_5, 124
Tone Synthesis variable resistor R_5 and R_6, 74, 76
Tools, 22, 23, 24
Touch alarm, 176-177
Transformer, symbol, 47
Transistor:
  bi-polar, 189
  symbols, 47
Transistor Q_1, 188
Triac, 47, 51, 52, 54, 100
Trigger Level Adjust R_3, 105, 173
T.V. Commercial Killer, 65, 70-71
"Tweeter" type speaker, 76
12-volt, high-current, battery-powered supply, 228
12-volt, low-current, 217-218
12-volt automatic battery charger, 138-140
12-volt sounder, 174
12-volt to 15-volt adjustable, high-current, 208-210
12-volt to 120-volt inverter, 137-138
2-conductor speaker wire, 30

Index

2-Way Intercom, 117, 119-120

U

Up/down control for solar array, 93-95

V

Variable-frequency square-wave generator, 190-192
Variable resistor, symbol, 45
Ventilation fan, 51, 58
Ventilation holes, 29, 38-39, 41
"Vernier" variable resistor R_3, 153
Vibrations, pick up, 149
Volume Adjust R_2, 118, 122, 124
Volume Adjust R_3, 121, 124, 189
Volume Adjust R_4, 151
Volume Adjust R_8, 158, 159
Volume Adjust R_9, 173

W

Warble Adjust variable resistor R_2, 68
Warning mechanisms:
  Alarm ON Time Adjust R_4, 180
  Alarm ON Time Adjust R_6, 174
  Arm Switch S_1, 176, 177
  attract attention when triggered, 163
  battery back-up unit, 163, 164
  break-beam, 164, 172-173
  building silent sentries, 163-164
  buzzer B_1, 165, 166, 172
  capacitor C_3, 173
  chemicals cabinet, 164-166
  complete home alarm system, 173-176
  continuous power supply, 163
  fire alarm, 167, 168
  general purpose timer, 179-181
  high-temperature, 164, 167-169
  high-water, 164, 166-167
  light-detecting, 178-179
  light striking photo-cell pickup, 164
  line-powered, 163, 164
  Lock Switch S_6, 174
  loud sounder, 163
  low-temperature, 164, 169-171
  medicine cabinet, 163, 164-166
  neighbors phone police, 163
  photo-cell R_1, 178
  photo-cell R_2, 172
  Push to Reload Switch S_1, 180
  reliability, 163

Warning mechanisms (cont.)
  second pair of "eyes," 164, 172-173
  Sensitivity Adjust R_1, 176, 177
  Sensitivity Adjust R_2, 167
  Sensitivity Adjust R_3, 178, 179
  switches S_a through S_d, 173
  switches S_1 through S_4, 173
  switch S_1, 165
  switch S_5, 174
  Temperature Adjust R_1, 168
  Temperature Adjust R_3, 167, 168, 169, 171
  temperature conditions, 163
  Thermistor R_1, 167, 168
  Thermistor R_2, 169, 171
  touch alarm, 176-177
  Trigger Adjust R_3, 173
  12-volt sounder, 174
  uninterrupted service, 163
  Volume Adjust R_9, 173
Water, high-, 164, 166-167
Water heater, 100-101
Wind-through-trees, sound, 117
Wire-wrap techniques, 34
Wiring for dependability, 30-31
Workbench:
  chair, 22
  comfort, 22, 23
  lighting, 22, 23
  oscilloscope, 24
  power supplies, 24
  quiet location, 22
  solid work surface, 22
  test instruments, 24
  tools, 22, 23, 24

Z

Zener diode, 20, 29, 42
Zener diode-regulated supplies:
  1.5-volt to 9-volt adjustable, low-current, 219-221
  5-volt, low current, 214-215
  9-volt, low current, 215-216
  ±9 volt, dual complementary, low current, 221-223
  12-volt, low-current, 217-218
  15-volt, low-current, 218-219
  ±15 volt, dual complementary, low-current, 223-225